AIR POLLUTION, OCCUPATIONAL SAFETY, & HEALTH, AND CLIMATE CHANGE:
FINDINGS, RESEARCH NEEDS, AND POLICY IMPLICATIONS

Establishing a GEO Health Hub Platform for Eastern Africa

University of Kabianga, Kenya
Great Lakes University of Kisumu, Kenya
Maasai Mara University, Kenya &
University of Southern California, USA

BLANK

Prepared by:

Augustine Afullo[1, 8], Christopher Onyango[2], George Ngatiri[2], Kiros Berhane[3], Jonathan Samet[3], Nuvjote Hundal[3], Francis Inganga[4], James Mwitari[5], Titus Oyoo[6] and Zablone Owiti[7, 8]

[1]Maasai Mara University, afullochilo2012@gmail.com;
[2]Great Lakes University of Kisumu, Kenya;
[3]University of Southern California (USC), USA;
[4]National Environment Management Authority (NEMA-Kenya);
[5]Ministry of Health, Kenya
[6]Kenya Bureau of Standards (KEBS), Kenya;
[7]National Commission for Science, Technology and Innovation (NACOSTI), Kenya; and
[8]University of Kabianga (UoK), Kenya.

Corresponding Author:

Prof Augustine Afullo,
Associate professor of Environmental Health,
University of Kabianga, Kapkatet Campus,
Department of Environmental Health,
P.O. BOX 2030-20200,
Kericho, Kenya
Email: afullochilo2012@gmail.com , augustine.afullo@fulbrightmail.org, a.afullo@kabianga.ac.ke; skype name: afullochilo2012
Mobile: +254722690956

ISBN: 9966-7205-9-6

Published by:

Wamra Technoprises, in collaboration with University of Kabianga, University of Southern California, and Great Lakes University of Kisumu. Nairobi, Kenya

Printed by:

GALLOP TECHNOLOGIES

P.O. BOX 22931-00400 Tom Mboya Street

Nairobi

Tel: + 254 733 499,736 833 358, 727 166414

EMAIL:

sales@galloptechnologies.com,admin@galloptechnologies.com

Citation:
A Afullo, C Onyango, G Ngatiri, K Berhane, J Samet, N Hundal, F Inganga, J Mwitari , T Oyoo and Z Owiti (2015). Situational Analysis and Needs Assessment in Kenya: Air Pollution, Occupational Safety And Health, And Climate Change for Health: Findings, Research Needs, And Policy Implications. Establishing a Global Environmental and Occupational Health (GEOHealth) Hub Platform for Eastern Africa. Wamra Technoprises, in collaboration with University of Kabianga (UoK), University of Southern California (USC), and Great Lakes University of Kisumu (GLUK), Nairobi, Kenya.

Acknowledgement:

The Lead author, on behalf of all participating / collaborating institutions, wish to extend a heartfelt gratitude to the National Institutes of Health (NIH), USA, for providing the Planning Grant funds (2012-15) for undertaking this situational analysis for Kenya. This came under the Global and Environmental Health (GEOHealth) Hub program 1R24TW009548-01.

Similarly, we wish to acknowledge and appreciate the participating institutions such as the Great Lakes University of Kisumu (GLUK) and other Universities which willingly gave information, the research and policy institutions which were sampled, as well as the national agencies such as the Kenya Bureau of Standards (KEBS), The National Environment Management Authority (NEMA), The Ministry of Health, and the National Commission for Science, Technology and Innovations (NACOSTI), without whose support this work could not have seen the light of the day.

Lastly, we wish to appreciate the Research Assistants who helped in the process of data collections, including Jentrine, Magdaline and Christopher.

Abstract

This situational analysis was undertaken in 2014 with the support of the National Institutes of Health (NIH, USA) planning grant of 2012/15. The aim was to assess the research training and policy needs and gaps in Occupational health and safety, climate change for health, and Air pollution for Health. This was necessitated by the increasing burden of disease associated with environmental and occupational exposures. At least 40 training, policy and research institutions were sampled across the country, and key informant interviews, as well as documentary and literature reviews conducted. The findings indicated an increasing burden of disease associated with climate change, air pollution and occupational health and safety. There were major gaps in training, research and policy, with local research contributing only less than 10% of all policies in the country. This was attributed to inconsistency and lack of continuity in collection of relevant data to form reliable data bases, rendering the researches not very credible. In addition, the limited research findings were delayed by a mean of 6 years before publication, at which time they have become stale. . On training, major gaps were found, with lack of basic competency and confidence among the graduates of the relevant programs in the thematic areas. This was attributed to lack of appropriate exposure to practical and practicums, as well as field trips which would enable tem relate the theories learnt to the life experience. Thus they largely required retraining. In the majority of the curricular, the four thematic areas were not mainstreamed, rendering them as areas lacking competent personnel. Whereas major gains have been made in policy frameworks, they were largely not effectively implemented, nor monitored. As a result, there is need for continuous collaborative research and training, coupled with policy interventions among universities, professors, students, research and policy institutions to enable the country make headway in air pollution for health, climate change for health and occupational health and safety. These needs would be sufficiently addressed by the formation of the proposed full GEOHealth Hub for Eastern Africa.

Table of Contents

Table of Contents

Table of Contents

Table of Contents

ABBREVIATIONS

AAP	Africa Adaptation Program
AAU	Addis Ababa University
ACPC	African Climate Policy Center
AMREF	The African Medical Research Foundation
CCAA	Climate Change Adaptation in Africa
CDM	Clean Development Mechanism
CDS	College of Development Studies
COMESA	Common Market for Eastern and Southern Africa
COTU	Central Organization of Trade Unions
COTU-K	Central Organization of Trade Unions (Kenya)
CUE	Commission for University Education
CUEA	Catholic University of East Africa
DOHSS	Directorate of Occupational Health and Safety Services
EACCCP	East African Community Climate Change Policy
EAPN	East Africa Pesticide Network
EEPA	Ethiopian Environmental Protection Authority
EMCA	Environmental Management and Coordination Act
EPA	Environmental Protection Agency of the United States of America
FANR:	Food, Agriculture and Natural Resources
FANRPAN	Food, Agriculture and Natural Resources Policy Analysis Network
FiTs	Feed in Tariffs
FKE:	Federation of Kenya Employers
GEF:	Global Environmental Facility
GEOHealth	Global Environmental and Occupational Health
GHGs:	Greenhouse Gases
GLUK	Great Lakes University of Kisumu
GOK	Government of Kenya
HOAREC/N	Horn of Africa Regional Environment Center and Network
HSEAC	Health, Safety and Environment Advisory Committee
ILO	International Labor Organization
ISO	International Organization for Standardization
ITDG	Intermediate Technology Development Group
IUF	International Union of Food
JKUAT	Jomo Kenyatta University of Agriculture and Technology
KACCAL	Kenya Adaptation to Climate Change in Arid and Semi-arid Lands
KARI:	Kenya Agricultural Research Institute
KEBS	Kenya Bureau of Standards
KEMRI	Kenya Medical Research Institute
KEMU	Kenya Methodist University
KEPHIS	Kenya Plant Health Inspectorate Service
KIPPRA	Kenya Institute of Policy Analysis
KIPPRA	Kenya Institute for Public Policy Research and Analysis
KIRDI	Kenya Industrial Research Development Institute
KMS	Kenya Meteorological Service

ABBREVIATIONS

KMTC	Kenya Medical Training College
KNUT	Kenya National Union of Teachers
KPAWU	Kenya Plantation and Agricultural Workers Union
KU	Kenyatta University
KUSPW	Kenya Union of Sugar Agricultural Plantation Workers
KWUST	Kiriri Women's University of Science and Technology
LMIC	Low Middle Income Country
MEAs	Multilateral Environmental Agreements (MEAs)
MKU	Mount Kenya University
MMU	Maasai Mara University
MMUST	Masinde Muliro University of Science and Technology
MOH	Ministry of Health Kenya
MOPHS	Ministry of Public Health and Sanitation
MU	Moi University
MUST	Meru University of Science and Technology
NACOSH	National Council for Occupational Safety and Health
NACOSTI	National Commission for Science, Technology and Innovation
NCCRS	National Climate Change Response Strategy
NDMA	National Drought Management Authority
NEAP	National Environment Action Plan
NEMA	National Environmental Management Authority
NGOs	Non-Governmental Organizations
NIE	National Implementing Entity
OSHA	Occupational Health and Safety Act
PCPB	Pest Control Products Board
PPE	Personal Protective Equipment
PSV	Public Service Vehicles
REC	Regional Economic Communities
RCE	Regional Center of Expertise for Education for Sustainable Development
SADC	Southern Africa Development Community
SANA	Situational and Needs Assessment
SEKU	South Eastern Kenya University
TUK	Technical University of Kenya
UNECA	United Nations Economic Commission for Africa
UNFCCC	United Nations Framework Convention on Climate Change
UoK	University of Kabianga
UoN	University of Nairobi
USC	University of Southern California
USI:	Unites States International University
WHA	World Health Assembly
WHO	World Health Organization
WIB	Work Injury Benefits Act

1.0 PROGRAM OVERVIEW AND OBJECTIVES OF KENYA'S SITUATIONAL ANALYSIS AND NEEDS ASSESSMENT (SANA)

1.1 Introduction

The Global Environmental and Occupational Health (GEOHealth) program supports paired consortia led by a low/middle income country institution (LMIC) and a United States (US) institution to plan research, research training, and curriculum development activities that address and inform priority policy issues concerning environmental and occupational health at the national and regional levels.

The burden of disease from environmental and occupational hazards and climate change is a growing concern in Eastern Africa, a region already facing challenges of malnutrition, poverty, and infectious diseases. To tackle these challenges, a comprehensive planning process involving information-gathering and synthesis, as well as networking, was undertaken by a multi-disciplinary team from the US, Ethiopia, and other Eastern African countries towards the eventual establishment of a GEOHealth Hub for Eastern Africa.

The GEOHealth Hub Platform for The East Africa-Ethiopia project, undertaken by the University of Southern California (USC) and Addis Ababa University (AAU), has the goal of taking crucial exploratory and planning steps toward establishing a hub for interdisciplinary research and training in environmental and occupational health for Eastern Africa. The focus is on three targeted areas:

- Air pollution and health
- Occupational health and safety
- Climate change and health

1.2 Phase I of the Project

Building on the rich and complementary experiences of USC and AAU in training and research activities in environmental and occupational health, this partnership was meant to strengthen the ongoing training, research, and policy-making activities related to environmental and occupational health and environmental sciences at the various schools of AAU, governmental agencies, and non-governmental organizations (NGOs) in Ethiopia, including the Ministry of Health (MOH), Ministry of Labor and Social Affairs (MOLSA), the Ethiopian Environmental Protection Authority (EEPA), and other Ethiopian Universities.

1.3 Phase II of the Project

This Situational and Needs Assessment (SANA) report was carried out in Kenya in response to the expansion of the hub to three other countries in Eastern African: Kenya, Rwanda, and Uganda. It involved evidence-gathering, evaluation of the current state of knowledge, identification of critical uncertainties, assessment of activities of key stakeholders, and an inventory of training, research, policy, and outreach capacity, targeting the three focus areas of indoor and outdoor air pollution and health, occupational health and safety (with focus on agriculture), and climate change and health. The targeted institutions were the Kenya Bureau of Standards (KEBS), Directorate of Occupational Health and Safety (DOSSH), National Environment Management Authority (NEMA), Ministry of Health (Research Division), Kenya Medical Research Institute (KEMRI), African Medical Research Foundation (AMREF), Great Lakes University of Kisumu (GLUK), Maasai Mara University (MMU), and other public and private universities in Kenya, among others listed under the methods section.

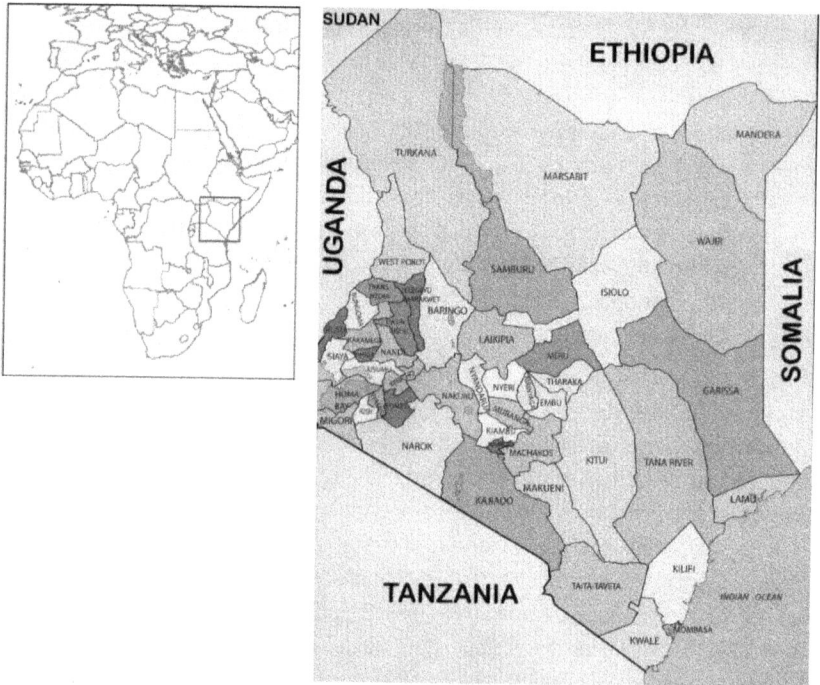

Figure 1: Kenya map showing Kenya's Counties

1.4 Objectives

The goal of the project was to establish the current state of the three thematic areas in Kenya and identify gaps in training, research, policy, and advocacy. The specific objectives were:

1. To establish the existing policies and acts on indoor and outdoor air pollution, occupational health, and climate change;
2. To evaluate the current level of training at institutions and universities on health and safety issues related to indoor and outdoor air pollution, occupational settings, and climate change;
3. To determine the type and level of research and research gaps on indoor and outdoor air pollution and health, occupational health and safety, and climate change and health.
4. To explore the advocacy approach on the focus areas of indoor and outdoor air pollution and health, occupational health and safety, and climate change and health.

1.5 Methodology

At least thirty key institutions were sampled for study based on their focus on the three targeted areas of concern; Indoor and outdoor air pollution and health, Climate Change and health, and Occupational health and safety. The sectors targeted were educational/training institutions, research institutions, policy institutions, and advocacy institutions. The regions covered were sampled from the Rift Valley, Western, Nyanza, and Nairobi regions (formerly provinces), constituting at least 24 counties of Kenya. The institutional and regional sampling was designed to ensure that the institutions accessed covered the national populace and had national representation covering the entire range of diversity in Kenya. Sampling was necessitated by the limited time and resource capacity to access all institutions, especially universities. Each region was covered by a trained research team who administered a questionnaire for Key Informant Interviews and a data collection tool designed for recording the information collected. In addition, an intensive literature review was conducted on all three thematic areas to establish the up-to-date status of data, their sources, and their use in development programs, e.g., the extent to which they would inform policy. While the bulk of the literature was online, some local university-based research documents identified as relevant were accessed directly.

For Training Institutions, the Research Assistant was expected to:

- Identify the appropriate universities;
- Visit, either physically or online;
- Review course advertisements for the last two months from archives and newspapers;
- List all relevant courses on offer for environmental and occupational health and the level offered;
- Check details of programs online to describe the lists of units for each course;
- Identify gaps if any and visit the universities to interview faculty in the relevant departments;
- Enquire about the major employers of graduates and the effectiveness of the graduates' university training;
- Sample the employers about the capabilities of graduates to help identify training gaps.

At the Research Level, the Research Assistant was expected to:

- Repeat the steps under the training level;
- Check all theses for master's programs relevant to environmental and occupational health, and abstract major theses;
- Take note of the profile, university, department, thesis, abstract, and the year completed;
- Review the website training on offer at the research institution, such as KEMRI and AMREF;
- Check the themes/activities of ongoing and past research;
- Pay an online or a physical visit to collect relevant abstracts

At Policy Level

The research assistants were assigned to check the institutional policy interventions, acts, and publications in any of the three thematic areas and identify institutional sources for the policies. Sampling was done systematically based on the following criteria: (i) regional coverage, with a focus on areas which have experienced the most disasters; (ii) communities most affected by some climate change related diseases, e.g., respiratory, resulting from use of a type of a housing unit; and (iii) national sectoral policy leadership. Next, the following institutions were sampled for study:

1. **Rift Valley area:** Egerton University, Moi University, University of Eldoret, Maasai Mara University, University of Kabianga, and Kabarak University;

2. **Western and Nyanza regions:** Maseno University, Masinde Muliro University, and Great Lakes University of Kisumu, as well as the East African Community and KEMRI;

3. **Nairobi and Central Kenya region:** Kenyatta University, Jomo Kenyatta University of Agriculture and Technology (JKUAT), University of Nairobi, KIPPRA, NEMA, KEBS, DOSHS, CUE, NACOSTI, AMREF, KEMRI, and KIRDI, as well as a number of advocacy institutions such as COTU, FKE and KNUT.

Since the policy institutions were national and largely based in Nairobi, rather than decentralized; the interviews were conducted in Nairobi

1.51 The Research Plan

Table 1 (in appendix) shows the various institutions visited for the Situational Analysis and Needs Assessment (SANA) for Kenya. The sampled institutions were grouped based on their area of focus and core business, geographical distribution, and thematic focus. The Rift Valley of Kenya had five institutions, all universities; the Western and Nyanza regions had five institutions of which two were policy-based while three were universities. Nairobi, as the capital city, had the majority: nine universities, ten policy institutions, and four advocacy institutions.

Table 2 (in appendix) further describes the exact information the research assistants were to seek from each institution of higher learning, either visited or studied.

The protocol was to secure an interview with ten graduates and ten employers of the graduates of the program and inquire about:

1. Where their employees were trained, and at what level of training;
2. General work experience of the employees;
3. Strengths of the graduates;
4. Weaknesses of the graduates;
5. Clear training gaps (areas in which the candidate showed most obvious deficiencies upon employment);
6. What the organization did to cope;
7. Cost implications implemented to cover for the deficiencies, e.g. further training or more intense supervision.

These same steps were to be carried out for policy and advocacy institutions and applied to the documents accessed. **Table 3** (in appendix) further describes the exact information the research assistants were to seek from each research institution visited or studied; while **Table 4** (in appendix) describes the exact information the research assistants were to seek for each research document accessed, such as theses at the institution visited or studied.

The exact information the research assistants were to seek for each policy document accessed and/or mentioned by those interviewed at the policy institutions is provided in **Table 5** (in appendix).

Name of project: Situational analysis and needs assessment (SANA) - Kenya.

Objectives of SANA:

- Identify the existing training, research, advocacy, and policy institutions and their operations
- Identify areas of need, gaps in research, policy, training, and advocacy

1.52. Target Institutions

- Learning institutions, e.g., universities
- Research institutions, e.g., KEMRI and AMREF
- Policy institutions, e.g., NEMA, DOSHS, KEBS, NACOSTI
- Advocacy institutions, e.g., RCEs

2.0 EXISTING LITERATURE ON THE GEOHEALTH THEMATIC AREAS

2.1 Introduction

Air pollution is an environmental and occupational health hazard, with pollutants in the form of either gas or particulate matter. The latter is a major component of outdoor air pollution and is widely used as a health-relevant indicator of air quality. Urban centers are likely to combine higher concentrations of atmospheric pollutants and a greater number of exposed people. Air pollution received recognition when the International Agency on Cancer Research (IARC) of the World Health Organization classified outdoor air pollution and particulate matter as carcinogenic to humans (WHO, 2013). Although the adverse effects of air pollution on human health have long been recognized, this new development is a major step in helping institutions mainstream pollution as a key target for action. After reviewing the latest scientific literature, IARC's panel of experts found sufficient evidence to conclude that exposure to outdoor air pollution is a major cause of cancer, i.e., carcinogenic.

2.2 Outdoor Air Pollution

By 2050, exposure to outdoor air pollution is projected to become the top environmental cause of premature death globally. This is more likely to be observed in developing countries where there is increasing industrialization and growth of urban areas combined with a lack of air quality standards. While many developed countries in North America and Europe have air monitoring programs, quality guidelines, and legislation aimed at mitigating pollution, this is not the case in many African countries. This fact received international recognition when, in 1992, the United Nations Conference on Environment and Development (UNCTED) made specific recommendations in its Agenda 21 (UN,1992) with regard to addressing air pollution in cities. One key recommendation was, "... the establishment of appropriate air quality management capabilities in large cities and the establishment of adequate environmental monitoring capabilities or surveillance of environmental quality and the health status of populations."

According to the US Environmental Protection Agency (EPA), Africa has some of the highest rates of population growth and urbanization in the world, with 38% of Africa's population living in urban areas, a figure estimated to reach 54% by 2030, and 60% by 2050.

Urban centers are likely to combine higher concentrations of atmospheric pollutants and an increase in the number of exposed people. Because of the increased urbanization in Africa, the rise in vehicle emissions and the trend towards greater industrialization, urban air quality in the continent is worsening. In many countries, the use of leaded gasoline is still widespread despite efforts to ban the practice, and vehicle emission controls are weak to nonexistent.

Kenya's potential to manage air quality improved tremendously with the passage of the Environmental Management and and Coordination Act (EMCA) of 1999 (Government of Kenya, 1999). Prior to the Act's passage, Nairobi lacked any regular air quality management system, and any measurements of air pollution had been carried out on an ad hoc basis. Indeed, out of 20 cities sampled from mainly developing countries for a UN study on air quality management capability, Nairobi's capacity received the lowest rating (UNEP/WHO, 1996, Odhiambo *et al.*, 2010, Thaddaeus *et al.*, 2013, and NEMA, 2009). There has, however, been some progress made toward air quality standards, although regulations drafted in 2009 remain in draft form and have yet to be formally gazetted for adoption.

Unlike industrialized countries where much research on motor vehicle air pollution has been conducted, little seems to have been carried out in Kenya. The small amount of information available on transport-related pollution in the city of Nairobi indicates high pollutant emissions (Gatebe, 1992; Gatebe *et al.*, 1996; Karue *et al.*, 1992, all cited in Odhiambo *et al.*, 2010)). A study by Odhiambo *et al.* (2010) on air pollution from motor vehicles in Nairobi found that air quality monitoring in most developing countries is not routinely conducted, and in some urban areas such information does not even exist, though signs of deteriorating air quality and health problems related to air pollution are visible. To assess air pollution levels in Nairobi and to develop appropriate air quality management plans for the city, it is necessary to have reliable information on the source(s) and the extent of the pollution.

Patrick *et al.* (2011) studied traffic-related $PM_{2.5}$ in the city of Nairobi in 2009. The mean daytime concentrations of $PM_{2.5}$ ranged from 10.7 $\mu g/m^3$ at the rural background site to 98.1 $\mu g/m^3$ on a sidewalk in the central business district. Horizontal dispersion measurements demonstrated a decrease in $PM_{2.5}$ concentration from 128.7 to 18.7 $\mu g/m^3$ over 100 meters downwind of a major intersection in Nairobi.

A vertical dispersion experiment revealed a decrease from 119.5 $\mu g/m^3$ at street level to 42.8 $\mu g/m^3$ on a third-floor rooftop in the central business district. Figures are average 0730-1830 hr weekday $PM_{2.5}$ concentrations near the Nairobi Fire Station (vertical dispersion site).

Patrick et al. (2011) further observed that motor vehicle traffic is an important source of particulate matter pollution in cities of the developing world, where rapid growth, coupled with a lack of effective transport and land use planning, may result in harmful levels of PM2.5 in the air. The lack of air monitoring data, however, hinders health impact assessments and the development of transportation and land use policies that could reduce health burdens due to outdoor air pollution. Establishing atmospheric monitoring programs is an essential step in tracking air pollution over a long time period. Such information would be invaluable in formulating policies to reduce emissions and in monitoring impacts on human health. Although few studies have examined the levels of particulate matter in urban areas across the continent, the available data indicate that there should be cause for concern.

In Nairobi, teams of students and researchers from New York City's Columbia University, the University of Nairobi, and Jomo Kenyatta University of Agriculture and Technology carried out a study on PM2.5 emissions from motor vehicles in 2009. By sampling along busy roads -- River, Ronald Ngala, Tom Mboya and Thika --, as well as in an open field, they sought to compare levels in high traffic areas and the outskirts of the city. The average levels reported in the journal Environmental Research and Policy ranged from 58 μg/m3 to 98 μg/m3 at the four sites in the city. These figures compare well with PM2.5 measured in other cities such as Cairo, (mean annual levels of 85 μg/m3); Accra (30-70 μg/m3); and Qalabotjha in Free State in South Africa (71-93 μg/m3). Current WHO guidelines for PM2.5 are an average of 25 μg/m3 for exposure over 24 hours, whereas the annual mean is 10 μg/m3. Because the sampling in Nairobi was done over 12 hours, it is difficult to make a direct comparison to the WHO guidelines. However, it is evident that particulate matter levels in African cities often exceed international guidelines.

Odhiambo et al. (2010) measured air pollutants extracted from air collected at the roundabout connecting University and Uhuru highways of Nairobi. This involved nitrogen oxides (NOx), ozone (O3), suspended particulates matter (PM10), and trace elements (lead or Pb). Sampling was done weekly for three months (February-April 2003), and measured hourly average concentrations. Measurements were carried out once per week for 8-hour durations from 0900 – 1700 hrs . Results showed that most pollutants -- notably, lead (0.051- 1.106 µg/m^3), bromine (LLD to 0.43 µg/m^3), NO_2 (0.011-0.976 ppm), NO (0.001-0.2628 ppm) and O_3 (LLD-0.1258 ppm) -- were within the WHO guidelines. However, PM_{10} levels (66.66 - 444.45 µg/m^3) exceeded WHO guidelines of 150 µg/m^3 for most days, with coarse particulate matter accounting for more than 70%. The study found a strong correlation (r = 0.966) between fine (0.4 µm) particulates, NOx, and motor vehicle density, which indicates the importance of traffic as a common source for both fine particulates and NOx. The PM_{10} levels were found to be higher than the recommended WHO guidelines; in some cases, they were more than 150% higher.

The study also found that the mean PM_{10} was 239±126 µg/m^3, with a range of 66.7-444.4 µg/m^3, with concentrations of coarse particulates increasing with little or no rain. In addition, a positive correlation was evident (r = 0.31) between vehicle density and coarse particulates. There was, however, a strong positive correlation between fine particles and motor vehicles density (r = 0.93). The study identified motor vehicle exhaust as the most probable source of fine particles. It also found a high correlation (r = 0.97) between fine particulates and NOx, indicating that fine particulates may have similar emission source(s) to NOx, a conclusion further supported by the strong correlation of NO and motor vehicle density (r = 0.94), and PM_{10} and motor vehicles density (r = 0.93) (Odhiambo et al. 2010).

Mulaku and Kariuki (2001) conducted a study on mapping and analysis of air pollution in Nairobi, and identified common air pollutants such as carbon monoxide, nitrogen oxide, sulphur dioxide, lead, and total suspended particulates (TSP) which include dust, smoke, pollen, and other solid particles. Cities and urban areas contain the bulk of people who are most vulnerable to the immediate effects of air pollution. The WHO recommends a maximum daily average TSP of 230 µg/m^3 and a maximum annual mean of 90 µg/m^3.

Ideally, every major city should have air quality management capability. This is defined as "... the capability to generate and utilize appropriate air quality information within a coherent administrative and legislative framework, to enable the rational management of air quality" (UNEP/WHO, 1996, cited in Mulaku and Kariuki, 2001). Most developing countries, including almost all African countries, have no air quality management capabilities despite having the fastest growing urban populations. One major consequence of the lack of air quality management capabilities in developing countries is the lack of data on air pollution.

The study by Mulaku and Kariuki specifically aimed at producing a basic spatial distribution map of TSP using GIS techniques. Data from 11 TSP air pollution sampling stations, which were distributed in commercial, industrial, and residential areas, were used. The sampling stations were positioned using hand-held GPS's and integrated with digitized base map data for Nairobi. Thiessen polygons were described around each sampling station to delineate surrounding areas, allowing for applicable measurements to be taken. These areas were then classified as having LOW (< 90 $\mu g/m^3$ annual mean), MEDIUM (90 – 180 $\mu g/m^3$), or HIGH (>180 $\mu g/m^3$) levels of TSP, and class boundaries were generated. A more accurate distribution pattern could be obtained by using a more complex dispersion model that would take into account the actual sources of pollution, distance from the source(s), meteorological conditions, topography, and census data. Most of these data, however, were not available for this study.

Kipkorir (2013) determined selected indoor and outdoor air pollutants in the central business district of Nairobi. The study compared indoor and outdoor levels of PM_{10}, heavy metals (Pb, Zn, Mn, Cu,) in PM_{10} and noxious gases (SO_2, NO, NO_2, and CO) in Nairobi's central business district from March-November 2007. PM_{10} was sampled using a Gent aerosol sampler and the heavy metals in the particulate matter were determined by an X-ray fluorescence technique. Gases were collected and analyzed using direct reading gas sensors. Statistical analysis of PM_{10} results showed that 53% indoor pollutant levels were higher than those corresponding to outdoor levels (I/O ratio >1.0). The mean average concentrations were 377.0±197.7 $\mu g/m^3$ and 280.1±122.5 $\mu g/m^3$ for indoor and outdoor pollution, respectively. This is attributed to fast mixing by turbulence outdoors, a situation which does not prevail in the still-air indoor environment. However, indoor/outdoor ratio analyses for heavy metals indicate higher concentrations in outdoor, compared to corresponding indoor, pollution (I/O ratio=0.9).

Ratios for noxious gases had indoor concentrations greater than outdoor concentrations, suggesting that indoor levels might have been influenced by indoor activities. Correlation analyses of PM_{10}, heavy metals, and gases between regions indicate significant positive and negative relationships (P > 0.050), suggesting differences in activities in the regions. Most concentration levels of PM_{10} and heavy metals surpass the set maximum concentration limits, while gases were within the limits.

2.3 Exposure Levels of Indoor Air Pollution

Health is influenced by a wide range of physical, social, and environmental factors. In addition to the production of toxic pollution, the supply and use of household energy in conditions of poverty and scarcity affects health – particularly the health of women and young children – in a variety of ways that encompass physical injury, lost opportunity for income generation, environmental stress, and many other issues. Indoor air pollution is the clearest and most direct physical health risk, and there is now fairly consistent evidence that biomass smoke exposure increases the risk of a range of common and serious diseases of both children and adults (Bruce *et al.*, 2000). The most important among these is childhood acute lower respiratory infection (ALRI), particularly pneumonia (Smith *et al.*, 2000 and Schirnding *et al.*, 2002). The illustration below (Martin *et al.*, 2013) shows a typical illustration of how different people in different geographical settings get sick through various exposures.

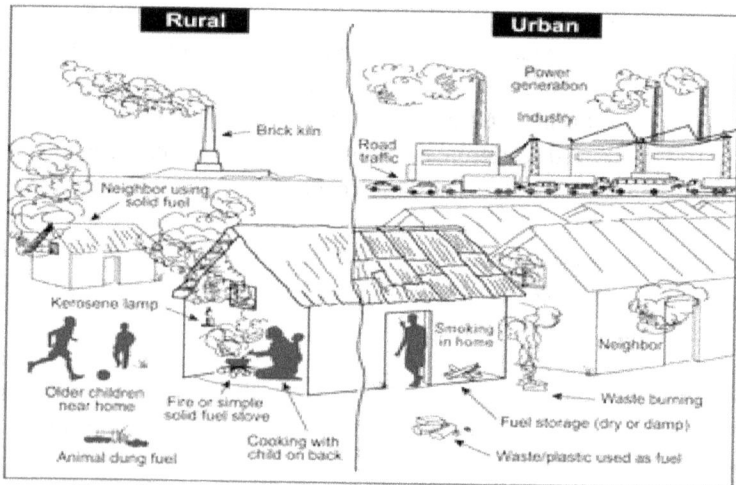

Figure 2: Illustration of various exposures to in-door air pollution by different people in different geographical settings (Source: Martin et al., 2013)

In poor rural and urban homes in Kenya, biomass fuels and coal are typically burnt indoors, with inadequate ventilation for the smoke. This leads to very high levels of pollution in the homes, with women and young children especially at risk for exposure. Smoke from these fuels contains many health-damaging pollutants, including particulate matter, carbon monoxide, nitrogen oxides, carcinogenic pyrenes, and benzene. Together, these pollutants irritate the airways and lungs, reducing the resistance to infection, and increasing the risk of cancer. Studies have measured particles which are thought to be the most health-damaging component of smoke pollution. This is significant, given that CO, NOx, formaldehyde, and polycyclic organic matter, including carcinogens such as benzo-pyrene, can damage health, with particulate matter able to accumulate to more than 100 times the threshold in such houses (WHO 2005, factsheet 292).

Particles under 10μ, but especially 2.5μ, in diameter can penetrate deeply into the lungs and appear to have the strongest potential for damaging health. Whereas it would be best to measure $PM_{2.5}$ to monitor indoor air pollution, such a process is highly technical, expensive, and difficult. As a result, particles up to 10μ in diameter (PM_{10}) have been most commonly measured (in $μg/m^3$ of air). Typical 24-hour mean levels of PM_{10} in homes using biofuels range from 300 to >3000 $μg/m^3$; during use of an open fire, the PM_{10} level can reach 20, 000 $μg/m^3$ or more. By comparison, the USEPA standard for daily (24-hour) average PM_{10} is 150 $μg/m^3$ (which should be exceeded only one in 100 days), while the annual average should not exceed 50 $μg/m^3$ (USEPA 1997). Most cities in the 'western' world rarely exceed these standards, whereas in rural homes in developing countries they are exceeded manifold on a daily basis. Levels of carbon monoxide and other pollutants also often exceed the standard guidelines.

The association between exposure to pollutants and chronic bronchitis and chronic obstructive lung disease and the fact that exposure to coal smoke in the home markedly increases the risk of lung cancer are well-documented in the peer-reviewed scientific literature (Schirnding et al., 2002). In recent years, further evidence has arisen indicating that indoor air pollution (IAP) in developing countries may also increase the risk of other important child and adult health problems (Schirnding *et al.*, 2002). The health and well-being of the mother and child are inextricably linked. Women normally continue their usual work during pregnancy, so the unborn child is exposed as the mother goes about her activities in the polluted kitchen. After birth, the young child typically stays very close to his or her mother until starting to walk; this means that the child is exposed directly, hence increasing the risk of a range of serious health problems.

Nearly 3 billion people worldwide rely on solid fuel combustion to meet basic household energy needs, with the resulting exposure to air pollution causing 4.5% of the global burden of disease. Clark et al. (2013) sought to identify research priorities for exposure assessment that would more accurately and precisely define exposure–response relationships of household air pollution necessary to inform future cleaner-burning cook stove programs. The following priority research areas were identified: improved characterization of spatial and temporal variability for studies examining both short and long-term health effects; development and validation of measurement technology and approaches to conduct complex exposure assessments in resource-limited settings with a large range of pollutant concentrations; and development and validation of biomarkers for estimating dose. Addressing these priority research areas will lead to better characterization of exposure–response relationships. Without improved understanding of these relationships, the level of air pollution reduction necessary to meet the health targets of cook stove interventions will remain uncertain (Clark et al., 2013).

The Bailis et al. (2005) study of Kenya Ceramic Jikos – a charcoal-burning stove used for cooking -- states that every year in developing countries, an estimated 1.6 million people die from exposure to stove smoke inside their homes. It is believed that this exposure constitutes 2.7% of the global burden of disease, with an estimated 396,000 deaths in Sub-Saharan Africa due to indoor smoke. Cooking with wood, dung, coal, and other solid fuels is a major risk factor for pneumonia among children and chronic respiratory disease among adults. Estimated exposure data from Kenya are as follows: 2,795 and 4,898 µg/m3 average daily exposure concentrations in young and adult women, respectively (Ezzati, Saleh, Kammen, 2000).

The Intermediate Technology Development Group (ITDG) Smoke and Health Project (ITDG 2002) on Reducing Indoor Air Pollution in Rural Households in Kenya aimed at contributing to the reduction of exposure to indoor air pollution. Two study areas were chosen: Kajiado, where ITDG was involved in the Maasai Housing Project, and two communities in West Kenya, where ITDG was engaged in the Stoves and Household Energy project. Baseline monitoring of the kitchens in these areas showed that smoke levels were very high: in Kajiado, the 24-hour average of respirable particulates was 5526 $\mu g/m^3$, while in West Kenya, the levels were 1713 $\mu g/m^3$. These values are much higher than the USEPA standards for acceptable annual levels of respirable particulates of 50 $\mu g/m^3$, indicating that the daily rates are over one hundred times greater in Kajiado and twenty times greater in West Kenya than the accepted values. These types of studies are useful for monitoring exposures, yet they are often difficult to measure and expensive to carry out. As a result they are rare.

WHO (2005) in its Household Air Pollution and Health Fact Sheet N°292 asserts that indoor air pollution from traditional cooking methods has serious health implications. High levels of wood smoke exposure have been reported in studies from many developing nations (Kammen et al., 1999). This, in turn, has been linked to acute respiratory infection (ARI), in particular pneumonia, as well as eye infections and burns (Ezzati et al., 1999). Without a substantial change in policy, the total number of people relying on solid fuels will remain largely unchanged by 2030 (World Bank, 2010). The use of polluting fuels also poses a major burden on sustainable development. Household air pollution (HAP) from solid fuel (biomass or coal) combustion is the leading environmental cause of death and disability in the world. Many governments, multinational companies, and non-governmental organizations are developing programs to promote access to improved stoves and clean fuels, but there is little demonstrated evidence of health benefits from most of these programs or technologies.

The Moturi (2010) study of risk factors for indoor air pollution in rural households in Mauche division, Molo District, found that burning of biomass fuels is among the major contributors to indoor air pollution (IAP). In Kenya, efforts to reduce IAP are largely institutional, leaving households unattended, where children under five years of age and their mothers bear the brunt of the adverse health impacts of using these fuels. The researchers in this study checked the type of primary buildings, number of rooms, type of ventilation present, and type of fuel used by the household. They found that 88% cook in primary buildings, 100% use wood fuels, the buildings have small windows with 15 cm open space between roof and wall (eaves space) serving as ventilation. Most houses had 30 cm^2 wooden windows, but 37% had no windows. In daytime, the doors remained open, making ventilation better. At night, however, ventilation was poor because doors were closed, with burning wood used for light, fuel, heating, or cooking, while people were in the house. However, health status was not monitored in this study: its purpose was to identify risk factors, not quantify their impacts. Particulate matter, chemicals, and infectious agents from open fires ravage the respiratory system of those living in such congested households.

The Moturi (2010) study) further found that 43% households had single-roomed – and therefore crowded -- houses, with at least 2 children present; these homes had little ventilation, with children spending most time indoors during cold and wet weather, as well as at lunch time when cooking is taking place and the children are exposed to smoke. ARI contributes 70% of deaths to "under-fives" in Kenya, and 4% of DALY, second only to malnutrition (16%) and WASH (9%). The Kammen-Ezzati (2010) study on indoor air pollution from biomass combustion and URI in Kenya, an exposure response study in Laikipia district, shows a strong association between IAP and disease.

The Intermediate Technology Development Group (ITDG) smoke and health project (1998) showed the positive health impacts of smoke hoods and eaves-spaces in a study in Kajiado, Kenya. The smoke hoods proved extremely effective, reducing the particulate levels from 4,383 µg/m3 to 1,075 µg/m3, while the carbon monoxide levels in the room fell from 48 ppm to 10.7 ppm. Increasing the size of the eaves spaces from small to large reduced the particulate levels from 2,042 µg/m3 in Kajiado to 766 µg/m3 in western Kenya. In Kajiado eaves spaces had not been adopted because it was difficult to cut them into the tightly-woven and mud-smeared walls close to the roof (ITDG, 1998).

ITDG's project of 'Traditional 3 stone cookers' assessed the impacts of energy technology on indoor air pollution. The Kenya study showed the impact of the various forms of household energy technology on the level of indoor air pollution resulting from smoke emission, and any improvements that may come about as an outcome of the introduction of improved, high-efficiency-low-emission ceramic cook stoves. Emissions under the conditions of operation by actual users were considered by conducting day-long (14-hour) monitoring of pollution levels at various points inside the house. Mean daily suspended particulate levels of 1000-5000 µg/m3 were found to be common among those households which used firewood, with maximum levels as high as 200,000 µg/m3. Results indicate that improved stoves (both wood and charcoal) reduce the average pollution level but there is still significant overlap between improved stoves and open fire.

Ezzati et al. (2000) found that suspended particulate matter and carbon emissions from biomass combustion are causally associated with incidence of respiratory and eye infections, and they suggest improved stoves as options for emission reduction. Ezzati et al. (2000) compared the emissions of suspended particulate matter and carbon monoxide from traditional and improved biofuel stoves in Kenya under the actual conditions of household use. Analysis of data from 137 14-hour days of continuous real-time emission concentration monitoring in a total of 38 households showed that improved (ceramic) wood-burning stoves reduce daily average suspended particulate matter concentration by 48% (1822 $\mu g/m^3$) during the active burning period and by 77% (1034 $\mu g/m^3$) during the smoldering phase. The greatest reduction in emission concentration was therefore achieved as a result of transition from wood to charcoal where mean emission concentrations dropped by 87% (3035 $\mu g/m^3$) during burning period and by 92% (1121 $\mu g/m^3$) when smoldering. These results indicate that a shift from raw organic fuels such as firewood, cow dung, and crop remains as energy sources to a more refined fuel such as charcoal, followed by the use of improved wood stoves, is a viable option for reduction of human exposure to indoor air pollution in many developing nations.

Ezzati et al. (2000a) observed that acute and chronic respiratory infections resulting from exposure to indoor particulate matter were the leading causes of morbidity and mortality in developing countries. Efforts to develop effective intervention strategies and detailed quantification of the exposure-response relationship for indoor suspended particulate matter require accurate estimates of exposure. This study used continuous monitoring of indoor air pollution and individual time-activity budget data to construct detailed profiles of exposure for 342 individuals in 55 households in rural Kenya. Young and adult women were found to have the highest absolute exposure to particulate matter from biomass combustion (4,016 $\mu g/m^3$ and 7616 $\mu g/m^3$ daily average exposure concentrations, respectively). Exposure during brief high-intensity emission "episodes" accounts for 38%-63% of the total exposure of household members who take part in cooking, and 0% - 15% for those who do not. Simple models that neglect the spatial distribution of pollution within the home, intense emission episodes, and activity patterns underestimate exposure by 35%-80% for various demographic sub-groups, resulting in inaccurate and biased estimations. Health and intervention impact studies should therefore consider the critical role of exposure patterns in detail, including the short periods of intense emission, to avoid spurious assessments of risks and benefits.

Ezzati et al. (1999) used individual exposure profiles and time-series health data to study the relationship between exposure to suspended particulate matter and the incidence of acute respiratory infection (ARI), eye infection, and headache. For adults, the study found significantly higher incidence of ARI ($p < 0.01$), eye infection ($p < 0.05$), and headache ($p < 0.01$) among females. Illness incidence rate, especially among adults, was significantly related to average exposure; a 1,000 µg/m3 increase in average TSP concentration would result in a 3% increase in the number of illness cases among adults and a 4% increase among children. This relationship disappears when gender is incorporated; this connection is attributed to cooking duties, performed entirely by women who are exposed to high concentrations within short periods of time. The intensity of exposure is as important as average exposure in determining health risks associated with indoor air pollution, implying that any further research on household fuel and health needs to factor in gender issues to establish the exact association with ARIs.

Kammen et al. (1999) integrated individual time-activity pattern data and continuous monitoring of indoor air quality to construct personal profiles for exposure to suspended particulates resulting from biofuel combustion in an array of stove-fuel combinations in rural Kenya. Analysis shows that on average the improved (ceramic) wood stoves reduce the level of suspended particulate emission by 50% from that of traditional open fire ($p = 0.07$). For individuals over 5 years of age, average personal exposure is correlated with gender, with women having higher exposure than men ($p = 0.01$). This was attributed to the extra time spent inside the house by female members of the household ($p = 0.01$), suggesting the need for a multiplicity of strategies for reduction of exposure to indoor air pollution, based on both reduction of emissions from stoves and modifications to the structure of the cooking area or the time spent near stove.

Saleh et al. (1999) observed that suspended particulate matter from combustion of biomass is causally associated with the incidence of respiratory and eye infections. The study integrated individual time-activity pattern and continuous monitoring of indoor air pollution to construct personal profiles for exposure to suspended particulates. This analysis shows that pollution levels vary considerably, even within small rural homes, with exposure highly influenced by the location and activities of household members, themselves determined by gender and age. The individual exposure profiles and time-series health data were used to study the relationship between exposure to suspended particulate matter and the incidence of acute respiratory infection (ARI), eye infection, and headache. It was inferred that there is a logarithmic relationship between exposure to suspended particulate matter and the probability of diagnosis with ARI, eye infection, or headache (p-values in the range 0.006 - 0.15 for different ailments). Each logarithmic increase in exposure results in an additional 1%, 1.2%, and 3% probability of being diagnosed with eye infection, ARI, and one of these three illnesses, respectively. The implication is that to achieve significant improvements in health, exposure and hence pollution levels must decrease considerably.

2.4 Baseline Status of Climate and Climate Change in Kenya

Analysis of the trends in temperature, rainfall, sea levels and extreme events points to clear evidence of climate change in Kenya. Studies indicate that temperatures have generally risen throughout the country, primarily near the large water bodies (GoK 2010). Other projections also indicate increases in mean annual temperature of 1 to 3.5°C by the 2050s (SEI 2009). This warming is leading to the depletion of glaciers on Mount Kenya (IPCC 2007, UNEP 2009).

Rainfall is also projected to increase with many models indicating an intensification of heavy rainfall especially during the wet seasons, and an associated flood risk. Seasonal rainfall trends are mixed, with some locations indicating increasing trends while others show no significant changes. The annual rainfall totals show either neutral or slightly decreasing trends due to a general decline in the main long rains (MAM) season. Specifically, global climate models (GCMs) predict rainfall increases in northern Kenya (by 40% by the end of the century), while a regional model suggests that there may be greater rainfall in the West; the rainfall seasonality, i.e., Short and Long Rains, are likely to remain the same. Rainfall events during the wet seasons will become more extreme by 2100, and, consequently, flood events are likely to increase in frequency and severity. Droughts are likely to occur with similar frequency as at present but to increase in severity.

The rising industrial emissions, use of charcoal and wood fuel and open burning of waste, are some of the main sources of atmospheric pollution, which also contributes to global warming. Figure 4 shows variability of average monthly ozone (O3) and carbon monoxide (CO) at the Mount Kenya Global Atmosphere (GAW) station. The ozone and carbon monoxide average values range between 20-50 ppb and 50-150 respectively. These patterns are consistent with the prevailing wind patterns and provide a unique opportunity to monitor sources, transport and sinks of air pollution over Kenya.

Global climate change affects human health through pathways of varying complexity and scale and with different timing (McMichael 2003). The impacts vary geographically depending on the environment, topography and vulnerability of the local population. Climate change and variability affect natural processes which in turn lead to an increased incidence of a range of diseases, such as asthma, malaria, diarrhea, Rift Valley Fever and nutrition-related ailments. Figure 3 illustrates the pathways through which climate change can affect human health.

Air pollution as a result of rising fossil fuel combustion to meet the energy needs of a growing population, has resulted in increased frequency of cardio-respiratory diseases such as asthma. This implies that there is synergy in mitigation efforts to address climate change through clean energy alternatives while at the same time addressing fossil fuel related pollution-related ailments (Figures 3 and 4 below)

Figure 3: Pathways by which climate change affects human health (courtesy of Patz, 2000)

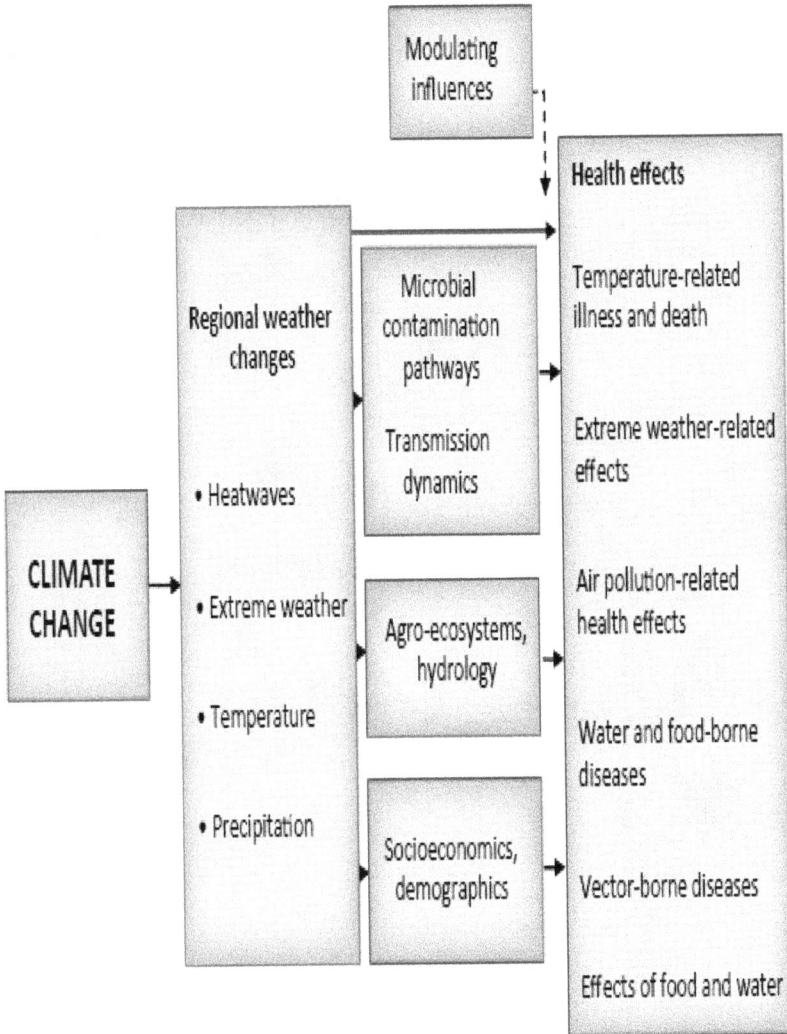

(a)

(b)

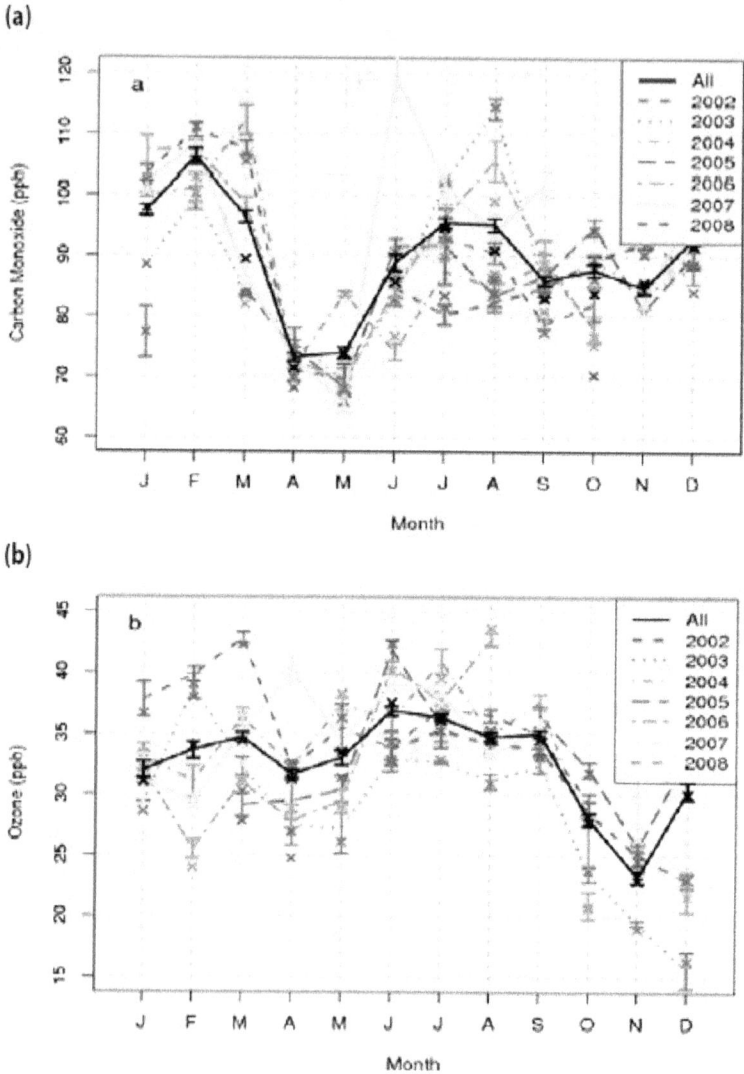

Figure 4: Climatology of (a) carbon monoxide and (b) ozone at the Mout Kenya Glabal Atmospheric Watch (GAW) station for the period 2002 – 2008 (courtesy of Kenya Meteorological Department)

2.41 Current state of Kenya Climate Change database, research, and monitoring

Agencies involved in expanding climate change knowledge in Kenya include government, research, and academic institutions, more than 300 civil society organizations (CSOs), and private sector companies. There is need for a standardized method of organizing this knowledge. Whereas various institutions have developed basic databases, libraries, and websites for storing climate change knowledge, there is limited sharing of such information and knowledge across government, the private sector, CSOs, academic and research institutions, and individual researchers. A few organizations are, however, starting to share their knowledge, but to date this has been on a small scale. An example is the Arid Lands Information Network (ALIN) which publishes *Joto Afrika*, a quarterly magazine that carries climate change research briefings by African scientists. In 2011, ALIN partnered with the Climate and Development Knowledge Network (CDKN) and CARE to develop special issues of *Joto Afrika* in the build-up to COP 17. There is a need to establish an electronic system for managing climate change related knowledge.

For greenhouse gas (GHG) data, the major challenge lies, in institutional arrangements necessary for effective compiling, archiving, up-dating, and managing inventory data as specified by the IPCC as good practice. Past projects on data collection in Kenya have been managed and spearheaded by the Ministry of Environment and NEMA, with overlapping mandates and responsibilities between them, with no clear delineation of authorities. They are supported by an ad hoc committee called the National Climate Change Activities Coordinating Committee (NCCACC). Because of this vagueness in the management of past climate change activities, institutional memory has been lost with the dissipation of national inventory experts. Data acquisition, especially from other line ministries and government agencies, as well as from the private sector, also seems to present a major problem. It is, however, hoped that these challenges will be addressed with the implementation of the NCCRS through the National Climate Change Action Plan (NCCAP) and enactment of the Climate Change Bill, 2014.

Many organizations dealing with climate change host large research repositories. Unfortunately, these repositories of information and knowledge are not interlinked to facilitate ease of sharing and wider access.

Examples of these repositories include databases, web portals, and libraries, as listed below.

- The University of Nairobi's Institute of Climate Change and Adaptation uses library services, as well as web-based and in-house databases.
- The Kenya Association of Manufacturers (KAM) uses a database documented as best practices
- The World Agroforestry Center (ICRAF) uses databases, library services, online scientific journals, web portals, and scientific content management sites.
- Other repositories include the Kenya Environment Information Network (KEIN), used by NEMA, as well as a large database of non-codified indigenous knowledge that communities have adopted as coping techniques in response to climate change effects.

Although the organizations are developing these repositories of information and knowledge, the next step of interlinking those repositories to make them widely accessible has yet to occur. Furthermore, the systems are likely to be incompatible or difficult to integrate because they have been developed on diverse platforms, therefore lacking common standards.

2.42 Climate Change in Kenya

According to NEMA, climate change is possibly the most significant environmental challenge of our time and it poses serious threats to sustainable development in Kenya. It impacts ecosystems, water resources, food, health, coastal zones, industrial activity, and human settlements. Addressing these impacts, however, presents opportunities for innovation, business, and improved livelihoods. Kenya lies along the equator and experiences wide climatic variation due to its physiographic diversity. The Inter-Tropical Convergence Zone (ITCZ) has considerable influence on the country's climate.

Unless effective mitigation and adaptation mechanisms are instituted with urgency, the combined effect of the impacts of climate change will slow or even hinder achievement of the targets detailed in Vision 2030. Vision 2030 is Kenya's strategic plan and route map towards becoming a middle-income industrialized country targeting the year 2030, and is the benchmark for all national programs and projects. It is therefore important to formulate a range of policy instruments to address climate change. While the National Climate Change Response Strategy (NCCRS) was finalized in 2010, it is necessary to go further and formulate a national policy on climate change and enact a climate change law.

An analysis of the trends in temperature, rainfall, sea levels, and extreme events points to clear evidence of climate change in Kenya. The country's arid and semi-arid lands (ASALs) have also witnessed a reduction in extreme cold temperature occurrences. This warming is leading to the depletion of glaciers on Mount Kenya. Because of the vital ecological role of mountains, this will have negative implications on biodiversity and water supply in the country as well as tourism, whose continued double-digit growth is crucial to achieving the 10% economic growth rate anticipated by Vision 2030.

The Government of Kenya (GoK) takes climate change and its impact on development seriously. Between 2005 and 2015, the GoK committed approximately KSh37 billion (USD 438 million equivalent) while development partners have committed KSh 194 billion (USD 2.29 billion equivalent) to programs that they classified as having a 'significant' or 'principal' climate change component (Government of Kenya, 2013).

Climate change is considered a cross-cutting issue that must be mainstreamed in all the sectors of the economy through the planning process. The Medium-Term Plan (2013-17), provided an opportunity to incorporate climate change programs into the national development plans, and build on both the National Climate Change Response Strategy and its Action Plan.

Figure 5: Forest cover of Kenya. It has increased from 2% five years ago to 6% current status.

2.43 Air pollution and the state of atmospheric environment in Kenya

The country's growing number of motor vehicles is commensurate with growth in the human population. Vehicles emit significant levels of air pollutants, including GHGs, while charcoal burning emits methane (CH_4), carbon monoxide (CO), and particulate matter into the atmosphere. These, together with rising industrial emissions, use of charcoal and wood fuel, and open burning of waste, are some of the main sources of atmospheric pollution.

Important sources of atmospheric contamination in Kenya include emissions from factories and motor vehicles, smoke from biofuels, and smoldering of copper cables and old tires. Kenya lacks consistent air quality monitoring stations, although some have been established on an ad hoc, demand-driven basis. An inventory on persistent organic pollutants, however, targeting dioxins and furans was undertaken recently and identified open burning of wastes as one of the largest sources of air pollutants in the country.

Some of the anticipated impacts of climate change in Eastern Africa include the following:

- decreased rainfall,
- rise in temperature and evaporation in dryland areas,
- higher frequency of drought spells,
- severe water shortage,
- change in planting seasons,
- reduced forest cover and arable land,
- increased outbreak of fungal attacks and insect infestations in agriculture due to changes in temperature and humidity,
- decline in crop yield and biomass production,
- increased risk of food shortages and famine,
- increased malaria transmission and national health care burden, and
- rise of sea levels within the coastal region.

One of the major challenges in environmental research is to identify appropriate adaptation mechanisms and coping options for the expected impacts of climate change.

The urban built environment in Kenya has a major impact on the overall environment. It is responsible for the loss and conversion of natural resources, especially forests and wetlands. It is also a major cause of environmental pollution, including the widespread problem of solid wastes. Cities and towns are major contaminants of the atmosphere through the increased use of fossil fuels in factories and by automobiles

2.5. Impacts of Climate Change and Air Pollution on Human Health

Kenya, like the rest of its counterparts in Eastern Africa, lives in interesting yet unpredictably challenging times due to climate change. There are more "unknowns" than "knowns" due in part to lack of proper database even for close monitoring and analysis of climate expectations and trends. However, because of the generic scientific knowledge about climate change, the following sub-section gives a broad overview of the socio-economic expectations resulting from climate change and air pollution, especially in relation to preparedness, health, and social protection.

Weather and climate play significant roles in people's health. Changes in climate affect the average weather conditions that are the norm for Kenya. Warmer average temperatures will likely lead to hotter days and more frequent and perhaps longer heat waves, which could increase the number of heat-related illnesses and deaths. Increases in the frequency or severity of extreme weather events such as storms could increase the risk of dangerous flooding, stronger winds, and other direct threats to people and property. Warmer temperatures could increase the concentrations of unhealthy air and water. Changes in temperature, precipitation patterns, and extreme events could enhance the spread of some diseases. The impacts of climate change on health will depend on many factors. These factors include the effectiveness of a community's public health and safety systems to address or prepare for the risk, and the behavior, age, gender, and economic status of individuals affected.

Impacts will likely vary by region, the sensitivity of populations, the extent and length of exposure to climate change impacts, and society's ability to adapt to change. In Kenya, this level of preparedness is near zero; the appearance of any disaster is likely to enact a heavy toll on citizens who have no social protection. In addition, the impacts of climate change on public health around the globe could have important consequences for Kenya as the country is a tourist destination. It may either lose tourists, thereby reducing its economic growth, and/or have more diseases entering the country through the tourism ports. Within its borders, more frequent and intense storms may lead to more floods, which may require more disaster relief, while declines in agriculture may increase food shortages.

2.51 Impacts from Heat Waves

Heat waves can lead to heat stroke and dehydration, and are the most common cause of weather-related deaths. Excessive heat is more likely to impact populations in the country's highlands where people are less prepared to cope with excessive temperatures. Young children, older adults, people with medical conditions, and the poor are more vulnerable than others to heat-related illness. Climate change will likely lead to more frequent, severe, and longer heat waves in the hot season, as well as less severe cold spells in the cold season (June-August). The impacts of future heat waves could be especially severe in large metropolitan areas such as Nairobi, Kisumu, Mombasa, Nakuru, and Eldoret, among others. Heat waves are also often accompanied by periods of stagnant air, leading to increases in air pollution and its associated health effects.

2.52 Climate Change, Human Health and Welfare

An analysis of the impacts of global climate change on human health and welfare indicate that:

- Many of the expected health effects are likely to fall mostly on the poor, the very old, the very young, the disabled, and the uninsured.
- Climate change will likely result in regional differences in national impacts, due not only to a regional pattern of changes in climate but also to regional variations in the distribution of sensitive populations and the ability of communities to adapt to climate changes.

Adaptation should begin now, starting with public health infrastructure. Individuals, communities, and government agencies can take steps to moderate the impacts of climate change on human health.

2.53 Impacts from Extreme Weather Events (drought and floods)

The frequency and intensity of extreme precipitation events is projected to increase in some locations in Kenya, as is the severity of tropical storms, with increases in wind speeds and rain. These extreme weather events could cause injuries and, in some cases, death. As with heat waves, the people most at risk include young children, older adults, people with medical conditions, and the poor. Extreme events can also indirectly threaten human health in a number of ways. Kenya is a direct victim because at least 80% of its land mass is traditionally classified as arid and semi-arid (ASAL). Increased drought means the economy shrinks, the citizens generate less income, and the disposable income for health support is likely to suffer. As such, extreme events of floods and drought can:

1. reduce the availability of fresh food and water;
2. interrupt communication, utility, and health care services;
3. contribute to carbon monoxide poisoning from portable electric generators used during and after storms;
4. increase stomach and intestinal illness among evacuees;
5. contribute to mental health impacts such as depression and post-traumatic stress disorder (PTSD).

Figure 3. Linkage between the Palmer Drought Severity Index (PDSI) and GDP growth, Kenya, 1975-1995

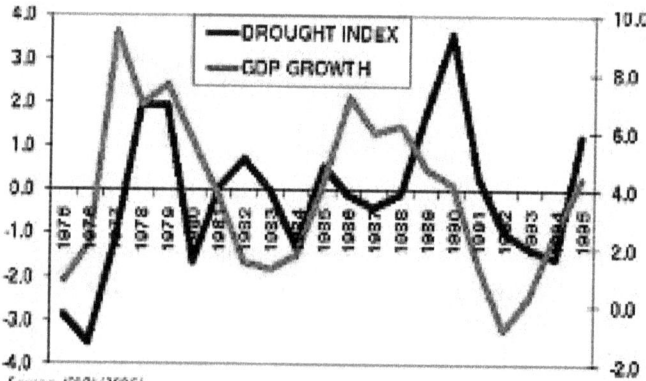

Source: IFPRI (2006).

Why climate change matters in Kenya: This figure shows the close relationship between drought events and GDP growth in Kenya over two decades (figure by IFPRI 2006). (Source) ttp://www.ilri.org/ilrinews/index.php/archives/6879

2.54 Impacts from Reduced Air Quality

In Kenya's cities and county headquarters, the citizens are exposed to increasing vehicular pollution. Now and in the future the residents are living in cities that do not meet air quality standards.

2.55 Increases in Ozone

Scientists project that warmer temperatures from climate change will increase the frequency of days with unhealthy levels of ground-level ozone, a harmful air pollutant and a component in smog. Concerns about increased exposure to ozone include the following:

- Ground-level ozone can damage lung tissue and reduce lung function, and inflame airways. This can increase respiratory symptoms and aggravate asthma or other lung diseases. It is especially harmful to children, older adults, outdoor workers, and those with asthma and other chronic lung diseases.
- Ozone exposure has been associated with increased susceptibility to respiratory infections, medication use, doctor visits, and emergency department visits and hospital admissions for individuals with lung disease. Some studies suggest that ozone may increase the risk of premature mortality, and possibly even the development of asthma.
- Ground-level ozone is formed when certain air pollutants, such as carbon monoxide, oxides of nitrogen (also called NO_x), and volatile organic compounds, are exposed to each other in sunlight. Ground-level ozone is one of the pollutants in smog.

Because warm, stagnant air tends to increase the formation of ozone, climate change is likely to increase levels of ground-level ozone in already-polluted areas such as in the cities of Nairobi, Mombasa, and Kisumu, and increase the number of days with poor air quality. If air pollutant emissions remain fixed at today's levels until 2050, warming from climate change alone could increase the number of dangerous exposure days.

2.56 Changes in Fine Particulate Matter

Particulate matter is the term for a category of extremely small particles and liquid droplets suspended in the atmosphere. Fine particles include particles smaller than 2.5 micrometers. These particles may be emitted directly or may be formed in the atmosphere from chemical reactions of gases such as sulphur, dioxide, nitrogen dioxide, and volatile organic compounds. In Kenya's cities, the large volume of vehicles is already heavily polluting the air, thereby directly exposing the population to respiratory diseases which records indicate are on the rise.

Inhaling fine particles can lead to a broad range of adverse health effects, including premature mortality, aggravation of cardiovascular and respiratory disease, development of chronic lung disease, exacerbation of asthma, and decreased lung function growth in children. Sources of fine particle pollution also include power plants, gasoline and diesel engines, wood combustion, high-temperature industrial processes such as smelters and steel mills, and forest fires.

Due to the variety of sources and components of fine particulate matter, it is clear that climate change will increase particulate matter concentrations across the country. Much particulate matter is cleaned from the air by rainfall, so increases in precipitation could have a beneficial effect. At the same time, other climate-related changes such as stagnant air episodes, wind patterns, emissions from vegetation, and the chemistry of atmospheric pollutants will likely affect particulate matter levels. Climate change will also affect particulates through changes in wildfires in Kenya's ASALS, which are also the ranching, livestock, and wildlife/ tourist resort sites. These are expected to become more frequent and intense in a warmer climate.

2.57 Changes in Allergens

Climate change may affect allergies and respiratory health, and the length of the allergy season may have increase. In addition, climate change may facilitate the spread of some invasive plants with highly allergenic pollen. This may affect Kenya's citizens in ASALS.

2.58 Air Quality and Climate Change

Improving Kenya's air quality is one of the top research intervention priorities. Recent studies have found out that:

- Climate change could increase surface-level ozone concentrations in areas where pollution levels are already high;
- Climate change could make national air quality management more difficult;
- Policy makers should consider the potential impacts of climate change on air quality when making air quality management decisions.

2.59 Impacts from Climate-Sensitive Diseases

Changes in climate may enhance the spread of some diseases. Disease-causing agents (pathogens) can be transmitted through food, water, and animals such as deer, birds, mice, and insects. Climate change could affect all of these transmitters.

2.60 Food-borne Diseases

Higher ambient temperatures can increase cases of salmonella and other bacteria-related food poisoning because bacteria grow more rapidly in warm environments. These diseases can cause gastrointestinal distress and, in severe cases, death. Flooding and heavy rainfall can cause overflows from sewage treatment plants into fresh water sources. Overflows could contaminate certain food crops with pathogen-containing feces. Already Kenya experiences annual flooding in some parts of the country, including Budalangi, Narok, Kano Plains and Baringo, often leading to serious health and social and economic impacts to the residents.

2.61 Water-borne Diseases

Kenya's disease records already indicate that water-related diseases such as water-scarce, waterborne, water-based and insect vector-borne diseases are on the rise. Heavy rainfall or flooding can increase water-borne parasites such as *Cryptosporidium* and *Giardia* that are sometimes found in drinking water. These parasites can cause gastrointestinal distress and, in severe cases, death. Heavy rainfall events cause stormwater runoff that may contaminate water bodies used for recreation with other bacteria. The most common illness contracted from contamination at beaches is gastroenteritis, an inflammation of the stomach and the intestines that can cause symptoms such as vomiting, headaches, and fever. Other minor illnesses include ear, eye, nose, and throat infections.

Mosquitoes favor warm, wet climates and can spread diseases such as West Nile virus. Various strains of West Nile virus prefer higher temperatures which could come with climate change; these are favorable to the survival of some existing/new strains. The geographic range of ticks that carry Lyme disease is limited by temperature. As air temperatures rise, the range of these ticks is likely to increase. Typical symptoms of Lyme disease include fever, headache, fatigue, and a characteristic skin rash. In addition, the malaria parasite is now found in literally all parts of the country.

Hitherto a preserve of warm lowland parts of the country, notably in Nyanza, Western, and Coast provinces, malaria now ravages the entire country, having climbed the hills and mountains to introduce a new form called highland malaria in the Kisii and Mount Kenya areas.

The spread of climate-sensitive diseases will depend on both climate and non-climate factors. Kenya has very weak public health infrastructure, lacking sufficient programs to monitor, manage, and prevent the spread of many diseases. The risks for climate-sensitive diseases can be much higher in the segment of the society living below poverty level (which comprises 50%) with less capacity to prevent and treat illness. Some need for social protection for them would be urgently needed, starting with appropriate research interventions.

2.62 Other Health Linkages

Other linkages exist between climate change and human health. For example, changes in temperature and precipitation, as well as droughts and floods, will likely affect agricultural yields and production. In some counties of the country, these impacts may compromise food security and threaten human health through malnutrition, the spread of infectious diseases, and food poisoning. The worst of these effects are projected to occur in arid and semi-arid areas, among vulnerable populations such as the elderly, vulnerable households, disabled, and the "under five"-age group. Declines in human health in other countries in the Eastern African region might affect Kenya through trade, migration, and immigration and have implications for national security.

Although the impacts of climate change have the potential to affect human health in Kenya and around the world, there is much that researchers, policy makers, and training institutions can do to prepare for and adapt to these changes. These should be intervention – based empirical research activities to develop an accessible database for climate change for health, air pollution and occupational health.

3.0 FINDINGS: ORGANIZATONAL / INSTITUTIONAL FRAMEWORK

3.1 Introduction

The findings are indicated according to the objectives of the situational analysis and assessment. This section discusses the organizational /institutional framework of the three themes of the GEOHealth project. The SANA was undertaken to establish the current status of, and gaps in, training, research, policy, and advocacy in air pollution, climate change, and agricultural occupational health and safety. A number of key institutions were visited during this process and information was gathered through Key Informant Interviews (KII) and desk reviews. Details regarding the institutions included in this section are further presented in **Appendix Table 6**.

3.2 The Nature of the Institutional Stakeholders in GEOHealth Themes

The policy and research organizations in Kenya are largely government parastatals created under specific acts of parliament. They are sectoral/thematic-based, with NEMA being in charge of environment; Bureau of Standards in charge of standards across all sectors; Directorate of Occupational Safety and Health (DOSH) in charge of occupational health issues; and universities undertaking research, policy, and training/capacity building at higher education levels. KIRDI's environmental management division also oversees research, technology, and innovation; it undertakes research and development, and technology transfer and training in the following areas:

- Sustainable industrial processes, including biotechnology, construction materials, leather and textiles;
- Efficient use of natural resources including renewable energy sources, biodiversity, minerals, and water;
- Waste management and pollution control in industry;
- Development of environmental management policies that integrate biophysical, social and economic systems, with industrial processes, with the aim to balance environmental conservation and economic development to provide for human well-being;
- Environmental impact assessments and audits for industrial and other projects; and
- Monitoring and evaluation, including environmental audits.

DOSH's research division has been enhancing its research capacity recently, and equipment acquired includes integrated sound level meters, indoor air quality monitors, a hematology analyzer, a biochemistry analyzer, and a laboratory incubator. OSH officers use occupational hygiene equipment for air sampling and noise measurements, and physicians and nurses use equipment in the medical laboratory for biological sampling and audiometric tests.

4.0: FINDINGS: TRAINING AND CAPACITY BUILDING ASPECTS OF GEOHEALTH

4.1 Introduction

This section presents the training and capacity building aspects of the GEOHealth project and the findings are indicated according to the objectives of the SANA.

4.2 The Training Institutions

4.21: The Non-University Training Institutions

Many institutions of higher learning, mainly universities, offer courses which relate to the three SANA themes. However, there are few specialized programs dealing with pollution and climate. The FKE also offers training in OSH, with trainers who are approved by the DOSHS. COTU trains its affiliated members, although the provision is based on the availability of funds. JKUAT offers both master's and postgraduate diploma courses in OSH, while other universities offering a master's degree in public health with a unit on OSH include Kenyatta University and Moi University. A few tertiary colleges offer diploma courses that have units in OSH, e.g., the Institute of Human Resource Management (IHRM). A number of key capacity building and training institutions visited include the universities and other institutions which directly offer services to the universities for reasons of quality assurance or collaborative research. These are presented in **Tables 7** and **8** below, with courses or higher education-related services offered by each institution clearly outlined.

Table 7: The Training profile and capacity building institutions:

Serial number	Institutions	Relevant training areas
1.	THE MINISTRY OF HEALTH	The Kenya Medical Training College (KMTC) is an institution, government Parastatal under the Ministry of Health in Kenya. It has at least 30 institutions called the medical training colleges (MTCs) These offers a post-basic diploma in OSH.
2.	AMREF (AFRICAN MEDICAL RESEARCH FOUNDATION)	AMREF Health Africa shares knowledge gained from its grassroots programs with others, and uses it as evidence to advocate appropriate change in health policy and practice. AMREF has a training program from MPH and nurses.
3.	DIRECTORATE OF OCCUPATIONAL SAFETY AND HEALTH SERVICES (DOSHS)	DOSHS works closely with some learning institutions such as Jomo Kenyatta University of Agriculture and Technology (JKUAT) to offer its short courses in OSH.
4.	CUE (COMMISSION OF UNIVERSITY EDUCATION)	CUE primarily deals with accreditation of programs such as those in climate change, pollution, and occupational health. Standards used to accredit programs are generic and apply across the board. The CUE aligns its quality management system with the requirements of ISO 9001:2008 statutory and regulatory requirements.

Approves curricula for various academic degree programs for institutions of higher learning; Grants charters to universities |

4.22 The Universities

The courses currently on offer in the various universities in all the three thematic areas are shown in **Table 8**, and are also examples of academic programs in environmental, occupational health, and climate change approved by the Commission of University Education (CUE) as of 2013; this was prior to the newly established universities having their programs fully available.

4.23 The Higher Education, Training and Capacity-Building Institutions on Climate Change, Air Pollution, and Occupational Health

Almost all institutions of higher learning in Kenya offer environmental and epidemiology-related courses. These include the University of Nairobi, Moi University, Jomo Kenyatta University, Egerton University, Kenyatta University, University of Kabianga, Maasai Mara University, and Maseno University, among others.

The private universities have varying areas of focus. For instance, Great Lakes University of Kisumu (GLUK) focuses on community health and development (a broader version of the master of Public Health course), though it has since diversified into other fields. More than fifty units are offered, encompassing environmental air pollution and climate change. GLUK has recently expanded: it has now stabilized / flattened, and is contending with how to handle the many graduate students registered. Of the over 50 research theses examined, only 3 (less than 10%) were within the air pollution sector.

Maasai Mara University (MMU) offers BSc, master's and PhD degrees in Environmental Studies, and has recently held a regional workshop on climate change, disaster risk reduction (DRR), and social protection (SP) in collaboration with the Organization for Social Research in Eastern and Southern Africa (OSSREA), which culminated in establishing a regional hub for climate change. This should offer excellent prospects for growth, given the partnership already in place from the various stakeholders.

Already MMU has developed a program/curriculum on occupational health at various levels (diploma, certificate, and degree), targeting all cadre stakeholders as well as the county government. The University of Nairobi, the oldest university in Kenya, has an institute for climate change and adaptation offering courses at the master s' and PhD levels.

But it has serious bureaucratic challenges which render access to its programs by outsiders a challenge, and at present it appears resistant to a paradigm shift. JKUAT has specialized in a master's program in occupational safety and health. Also at the master's level, science in environmental legislation and management is offered at the Institute of Energy and Environmental Technology.

4.3 Strengths, Weaknesses, Opportunities, and Threats (SWOT): Education and Training Analysis

4.31 The situation: Weaknesses and Strengths of Graduates

4.311 Strength of the training programs and their graduates
It has been asserted that fresh graduates from universities have such strengths as good technical knowledge, good research skills, and excellent work performance, that they initiate new projects, work under minimal supervision, and have specialized training, along with a willingness to learn.

4.312 Gaps and weaknesses in training
Weaknesses identified among the graduates of the various academic programs relevant to GEOHealth include poor problem-solving skills, poor time management, poor innovative skills, lack of exposure to practicals, lack of positive attitudes in their areas of specialization, and levels of specialization are not well-covered. An example of this last item is graduates with a B.Sc. Environmental Health and an Information Technology (IT) concentration have taken more units in IT than in environmental health. There was also evidence of poor work attendance. On the whole, training is largely theoretical, leaving the student/graduate with very little ability to tackle practical, real-world problems. Much of the learning appears to be rote, rendering graduates less than prepared for the job market.

4.32 Training Needs and Employer-Coping Mechanisms
As a result of these weaknesses, the employers or organizations need to provide more on-the-job training and strictly supervise the employees, as well as deploy personnel in particular areas of specialization. In the short run, these weaknesses can be addressed at the training institutes, with certain cost implications. The following suggestions could help to strengthen the training: 1) Learning institutions should expose students to more practical experience, with

more internships and attachments, engagement in field research, and report writing; 2) Students should be exposed to research paper presentations through cooperative learning, along with strict supervision; and 3) The learning institutions should expand their facilities to enable them to enroll more students. These steps could help reduce the amount of the rote learning currently taking place, which has limited the competency and awareness of many of the graduates. In the long run, there is a need to introduce a multi-sectoral intervention research agenda which involves the policy makers, the universities, and the communities to enable the universities to be local areas of excellence poised to find solutions to day-to-day common community/societal problems. The following is a checklist of some of the training needs.

4.33 The Training Needs Checklist

a) Establish laboratories for active student learning, as well as introduce field trips, attachments, and internships to better prepare the graduate for work;

b) Initiate more practical training;

c) Add more opportunities for attachments and internships where they already exist;

d) Initiate more effective exchanges of ideas between students and professors to improve the students' exposure and confidence;

e) Follow the sequence of training units correctly;

f) Ensure that the programs' units are relevant to their focus;

g) Ensure that the course title corresponds to the intended syllabus;

h) Introduce multi-disciplinary courses on climate change.

5.0: FINDINGS: THE RESEARCH ROLE OF INSTITUTIONS SPECIALISED ON GEOHEALTH THEMATIC AREAS

5.1 Introduction

The Situational Analysis and Needs Assessment was undertaken to establish the situation of, and gaps in training, research, policy an advocacy in air pollution, climate change, and occupational health and safety. This section focuses on research. A number of key research institutions visited include the Universities (public and private), NEMA, NACOSTI, CUE, KIRDI, and KEBS, among others. These are described in **Table 9** (in appendix).

5.2 Institutional Research Roles and Profiles

5.21 The Universities

Universities conduct research on all three thematic areas: climate change, occupational health, and air pollution. The majority of the research is done by graduate students for thesis and dissertation writing on a self-support basis. However, most self-sponsored research exhibits methodological, temporal, and spatial limitations. In many cases, lack of resources restrict the researcher to second or third option methods and equipment, rendering the majority of research not technologically up to date. As such, the majority of findings from locally/self-sponsored research never inform policy. Some research is conducted by faculty supported by local institutions such as NACOSTI and LVEMP. Most research supported by foreign agencies is time-bound, but again limited by spatial and temporal challenges. Results of foreign-supported research are largely unavailable to potential researchers, rendering duplication possible.

5.22 National Commission for Science, Technology and Innovation (NACOSTI)

NACOSTI seeks to promote and coordinated Research and Development (R & D) activities that aim to provide a clean, secure, and sustainable environment by (i) mainstreaming the Environmental Management and Coordination Act (EMCA) and National Environmental Research Agenda in all sectors; (ii) developing and maintaining an inventory on environmental R&D; (iii) enhancing collection, analysis, and access to environmental R&D data; and (iv) strengthening the linkages among industry, universities, and research institutions in pursuit of a curriculum and innovative products that support the sustainable management of the environment. As such, research is one of its key mandates. NACOSTI liaises with the Kenya the newly established National Innovation Agency and the National Research Fund to ensure funding and implementation of prioritized research programs, as well as coordinates annual research and innovation projects by theme. The latest theme has been heavily biased towards agriculture and food security. However, its scope is broad and it is a very relevant institution within GEOHealth.

5.23 Directorate of Occupational Safety and Health Services (DOHSS)

DOHSS has a key mandate in OSH and has continued to increase its research capacity. As mentioned, some of its equipment acquired recently includes integrated sound level meters, indoor air quality monitors, a hematology analyzer, a biochemistry analyzer, and a laboratory incubator. OSH officers use occupational hygiene equipment for air sampling and noise measurements, while physicians and nurses use the equipment in the medical laboratory for biological sampling and audiometric tests. Challenges include a lack of consistency in the records, complicated rights of access, issues of sustainability, and issues of reliability arising from methodology. As such there are no reliable data from DOHSS to assist in establishing a link between exposure and health.

5.24 Kenya Institute for Public Policy Research and Analysis (KIPPRA)

KIPPRA handles macro-economic and general national polices with an agricultural focus. As such it is not directly inclined towards GEOHealth. KIRDI has a high research profile, and works closely with universities in this respect. Its research focus is broad, with clean energy alternatives from waste, clean technologies (CDM), occupational health and safety, waste management, and carbon footprints (climate change mitigation),

5.25 The Commission of University Education (CUE)

The Commission of University Education (CUE) has a mandate to ensure the maintenance of standards, quality, and relevance in all aspects of university education, training, and research, which are the core businesses of universities. In addition, it coordinates the universities' government-funded research projects.

5.3 The State of Environmental Research in Kenya

Analysis of environmental research in Kenya (1967-2007) shows no significant increment of investment in environmental research after the enactment of the Environmental Management and Coordination Act (EMCA. 1999). The analysis of research institutions engaged in environmental research in Kenya showed that both UoN and Moi University were dominant in undertaking environmental studies between 1967 and 2007. The trend was reversed after the early 1990s, especially following the initial establishment of additional universities in the country such as KU and JKUAT, which also introduced strong environmental programs. Foreign universities and foundations had a strong lead in environmental research, accounting for 40.4% of all environmental research studies registered and permitted by the GOK between 1967 and 2007. UoN accounted for 13.75 %, while for MOI the figure was 9.9%. It is most likely that the involvement of foreign researchers has resulted in the movement of large amounts of scientific information and perhaps environmental specimens beyond Kenya, especially to Europe and North America. The EMCA (1999), has, however, through its institution, NEMA, developed a system of environmental assessment in which any newly proposed projects undergo a mandatory environmental impact assessment (EIA) while the existing projects undergo a mandatory environmental audit (EA). This has significantly increased the number of environmental studies in the country since 2004 when the registration of lead experts and agencies was formalized. However, this information remains only in NEMA records and libraries, with quarterly NEMA reports available only in its own libraries and those of several other subscribing institutions. Many relevant institutions are not aware of the existence of the reports. The analysis on areas of environmental research focus found the following:

o   Between 1967 and 1980, environmental research in Kenya was characterized by a strong focus on the biological environment, especially wildlife and issues on the human environment. Between 1981 and 2007, more research was conducted on the human environment in comparison to the physical and biological environments.

o   Moi University is the lead institution on research on the physical environment but with a greater focus on the aquatic environment.

o   UoN has conducted more research on the land environment than other institutions.

o   Most research on biodiversity and ecosystems is conducted by foreign scientists

5.4 Research and Research Capacity: Weaknesses

a)   Lack of practical exposure to research by students while studying;

b)   Lack of exchange between students and professors in research projects;

c)   Lack of collaborative research among a wide range of stakeholders so as to enable research to inform policy;

d)   Limited relevant research units in selected courses in undergraduate and graduate programs;

e)   Policy gaps resulting from lack of local data from empirical research in certain aspects of the environment;

f)   Research data in specific aspects of the environment are missing, leading to over- reliance on international data sources, or data from research conducted outside of Kenya, with little local relevance; Data have methodological, temporal and spatial limitations due to lack of time and resources, often even when the right approaches are known and the limitations are clearly identified;

g)   Use of data sources which are inadequate establish trends.

5.5 Research and Research Capacity: Needs

a)   More practical exposure to research by students while studying, e.g., as research assistants in faculty-coordinated projects;

b)   More attachment and internship opportunities at research institutions, or any involvement in research in which an intern or attaché can participate;

c)   Students and professor exchanges in research projects to improve research exposure and confidence of both, e.g., through co-supervision of student research by professors from different countries or institutions, with the possibility of field visits during data collection;

d)   Collaborative research among students, the community, researchers, and policy makers to enable trainees to acquire early grounding in research geared toward addressing contemporary societal problems, which then would render the research better placed to inform policy;

e) More relevant research units in systematically selected courses in undergraduate and graduate programs;

f) Use of policy gaps as sources for informing joint research by students, professors, policy makers, and the community;

g) Proper conduct of literature reviews to identify areas most in need of research, and the consideration of research-constrained areas;

h) Close partnerships in research conducted by foreign agencies, arising from a deep rooted needs assessment and situational analysis, coupled with a well-established MOU mechanism to guarantee ownership, full participation, and sustainability;

i) Long-term data collection mechanisms for various aspects of the environment, e.g., climate change, air pollution (indoor and outdoor), and health monitoring systems for various demographics in specific geographical and socio-economic settings;

j) Well-designed research plans, coupled with a proper information management system to facilitate long-term access and utilization by interested stakeholders. These should have a student component at various levels, e.g. MSc and PhD candidates, to give the upcoming researchers opportunities to not only develop themselves, but also to provide their best ideas and make a contribution to society. Alternatively, a student research grant component could be included in collaborative projects.

6.0: FINDINGS: POLICY ASPECTS OF GEOHEALTH

6.1 Introduction

This section deals with the results of the institutional and policy portion of the analysis. Among the key policy institutions relevant to the themes of GEOHealth are the NEMA, DOHSS, KEBS, and NACOSTI, with DOHSS as the national body having regulatory responsibilities in environment and exposure monitoring, medical examination, surveillance of workers' health, and advisory services. Other agencies include the National Environment Management Authority (NEMA), the former Ministry of Public Health and Sanitation (now Ministry of Health), and the former public health departments in the local authorities (now counties).

The Occupational Health Division in DOSHS undertakes occupational health surveillance in workplaces. It also monitors and supervises the activities of the designated health practitioners who carry out medical examination of workers. The Directorate's occupational hygiene and occupational health divisions are responsible for analytical and assessments related to determining workers' exposure to various occupational hazards.

Standards can be used to monitor and measure air quality and the country's carbon footprint. The Kenya Bureau of Standards (KEBS) has published a number of Kenya's standards in this regard, covering the environmental sector. These include: KS 1966-2 standard on tolerance limits for effluents discharged into public sewers; KS 1966-1 on tolerance limits for effluents discharged into surface waters; and KS ISO 14024 standard on environmental labels and declarations, as well as others.

6.2 Existing Policies, Legislations, and Related Thematic Frameworks

This section focuses on existing policies, gaps, and needs. Key among them are EMCA (1999), OSH (2007), WIBA (2007), and KEBS (2012-14).

6.21. Climate Change

An analysis of the trends in temperature, rainfall, sea levels, and extreme events points to clear evidence of climate change in Kenya. Because of the vital ecological role of mountains, climate change will have negative implications on biodiversity and water supply in the country and on tourism, whose continued double-digit growth is crucial to achieving the 10% economic growth rate anticipated by Vision 2030.

A number of institutions have been created by the existing policies and acts of parliament. Most of the instruments for environmental governance in Kenya have evolved from important global fora such as the Stockholm Conference on Human Environment of 1972 in Sweden, and the UN Conference on Environment and Development (UNCED) of 1992 in Rio de Janeiro, Brazil, and in Johannesburg in 2002. The government of Kenya has formulated a wide range of policies for sustainable development and environmental conservation which include:

(i)   Environment: Session Paper No. 6 of 1999 Entitled "Environment and Development," Environment Management Coordinating Act (EMCA), 1999 and Draft Environmental Policy.

This law provides for the relevant institutional framework to coordinate environmental management, including the establishment of the National Environment Management Authority (NEMA), which is the Designated National Authority (DNA) for Clean Development Mechanism (CDM) and the National Implementing Entity (NIE) for the Adaptation Fund. This landmark policy was enacted in 1999 by the Kenya Government and led to the formation of the National Environment Management Authority in 2003. Under EMCA, NEMA has the mandate of ensuring overall coordination, planning, regulation, and enforcement of environmental standards and overall compliance with the Act. The Ministry of Environment and Mineral Resources (MEMR) embarked on formulating an environmental policy, which is currently in draft form. The guiding principles in the policy include that of sustainable use and the 'Polluter and User Pays' principle.

(ii)   Water: National Water Policy and Water Act 2002: The EMCA 1999 and the Water Act of 2002 provide the overall governance of the Water Sector. The regulations and strategies established following this Act recognize the climate change implications on health, sanitation, and water.

(iii)   Wetlands: Draft National Policy on Wetlands Conservation and Management

(iv)   Forests: National Forestry Policy and Forests Act, 2005, and the Kenya Forestry Master Plan 1995-2020: The master plan provides an overarching framework for forestry development in the country for the 25 year period up to 2020 and served as the blue print for reforms in this sector, including the Forest Act of 2005 and Forest Policy of 2007. Together they recognize the environmental role of forests, including water values, biodiversity values, climate change values through carbon sequestration, and other environmental services.

(v) **Disaster: National Disaster Management Policy**: This legal framework institutionalizes disaster management and mainstreams disaster risk reduction in the country's development initiatives. The policy aims to increase and sustain resilience of vulnerable communities to hazards.

Multilateral Environmental Agreements (MEAs)

At the international level, Kenya is party to a wide range of global and regional MEAs such as:

(i) Convention for the Protection of the Ozone Layer (The Vienna Convention) and Montreal Protocol;

The convention is often called a framework convention, because it serves as a framework for efforts to protect the earth's ozone layer. The Vienna Convention was adopted in 1985 and entered into force on 22 Sep 1988. The objectives of the Convention were for Parties to promote cooperation by means of systematic observations, research and information exchange on the effects of human activities on the ozone layer and to adopt legislative or administrative measures against activities likely to have adverse effects on the ozone layer. The convention was to be implemented through the Montreal Protocol and at national level through country specific action plans. Kenya ratified the Montreal Protocol on 9 November 1988, the London and Copenhagen Amendments on 27 September 1994, and the Montreal Amendment on 12 July 2000

(ii) United Nations Framework Convention on Climate Change (UNFCCC)-Kyoto Protocol;

The international political response to climate change began with the adoption of the United Nations Framework Convention on Climate Change (UNFCCC) in 1992. The convention sets out the framework for actions aimed at stabilizing the atmospheric concentration of GHG at a level that would prevent dangerous anthropogenic interference with the climate system. The Conference of Parties, which manages the Convention and meets annually, adopted the Kyoto Protocol in 1997 (and entered into force in 2005) that commits industrialized countries and countries in transition to market economies to reduce their overall emissions of GHGs. Similarly, the convention requires all countries to take up climate actions taking into account their common but differentiated responsibilities and respective capabilities. Kenya ratified the convention 30th August 1994, thereby signifying her determination to join the international community in combating the problems related to climate change

Developing countries, including Kenya, are required to undertake Nationally Appropriate Mitigation Actions (NAMAs) in the context of sustainable development, supported and enabled by technology, financing, and capacity building. The voluntary implementation of NAMAs is aimed at achieving a deviation in emissions relative to 'business as usual' emissions by 2020. There is also a required system of measurement, reporting, and verification (MRV) and submission of biennial reports and national communications (every four years) by all countries; Kenya is in the process of preparing the 2nd Communication. Additionally, countries are expected to prepare and implement National Adaptation Plans (NAPs).

(viii)   Convention on Biological Diversity (CBD)

At the 1992 Earth Summit in Rio de Janeiro, world leaders agreed on a comprehensive strategy for meeting human needs while ensuring that they leave a healthy and viable world for future generations or "Sustainable Development". One of the key agreements adopted at Rio was the Convention on Biological Diversity.

The agreed text of the Convention on Biological Diversity (CBD) was adopted by 101 Governments in Nairobi, Kenya in May 1992 and signed by 159 Governments and the European Union at the United Nations Conference on Environment and Development (UNCED) held in Rio de Janeiro in June 1992. The Convention entered into force on 29th December 1993 and the first meeting of the Conference of Parties was held in Nassau, Bahamas, in November - December 1994.

Kenya is a mega bio-diverse country with over 35,000 species of flora and fauna including micro-organisms. These resources are of considerable domestic and intrinsic value. This diversity is a manifestation of the array of different habitats inherent in the major vegetation eco-zones, with species diversity dominated by insects. This diversity is served by the variable ecosystems ranging from marine, mountains, tropical, dry lands, forests and arid lands.

6.22 The situation: Climate Change and the State of the Atmospheric Environment in Kenya

The Environmental Management and Conservation Act (1999) does not address climate change, and there is little reference to the topic in Vision 2030. As such, unless effective mitigation and adaptation mechanisms are urgently instituted, the combined effect of the climate change-induced impacts will slow or even hinder achievement of the targets detailed in Vision 2030 (NEMA, 2012). The EMCA (1999) and the National Environmental Research Agenda provide adequate guidelines on environmental research and development. The function of the NACOSTI in relation to research and development is to develop and enforce codes, guidelines, and regulations in accordance with the policy determined under this Act for the governance, management, and maintenance of standards and quality in research systems.

6.23 Standards on Climate Change

Standards can be used to monitor and measure air quality, climate change, and carbon footprint. ISO has already developed standards with an impact on climate change for areas such as building environment design, energy efficiency of buildings and sustainability in building construction, intelligent transport systems, solar energy, wind turbines, nuclear energy, and hydrogen technologies. Such greenhouse gas (GHG) initiatives rely on quantifying, monitoring, reporting, and verifying GHG emissions and/or removals. The ISO standards directly related to climate change include ISO 14067, now under development, for measuring the carbon footprint of products (CFP). It will complement the already published standards ISO 14064 and ISO 14065, which provide an internationally agreed framework for measuring GHG emissions, verifying claims made about them, and accrediting the bodies which carry out such activities. KEBS has published standards on energy-saving lamps, covering performance and safety aspects. This includes KS IEC 60969, Kenya standard for the performance of self-ballasted lamps for general lighting services. This is because renewable energy and energy efficient technologies offer the benefit of not only mitigating climate change but also lowering energy costs. Increasing energy efficiency of appliances and equipment in residential, commercial, and industrial sectors will reduce energy-related carbon dioxide (CO_2) emissions.

6.3 Gaps in Climate Change Policies

Mainstreaming climate issues into the Kenyan government's programs is important, given that climate is a major driver of economic activities—such as agriculture and tourism—that are central to the achievement of Vision 2030. Currently, there are no policy-level instruments that deal specifically with climate change. In addition, whereas guidelines on environmental research and development exist, these have not been mainstreamed by relevant sectors. Another gap is community awareness about CFP, the community's role in containing the societal contribution to climate change, and the lack of CFP declaration by the various companies in the country, as well as the lack of related legislation to enable entities to compete based on their positive role in enhancing environmental quality.

6.4 Needs in Climate Change Policies

Policy strategies and instruments are thus needed to build climate change mitigation and adaptation mechanisms into the national development agenda. Besides formulating a national climate change policy and enacting a climate change law that would consolidate the gains made by the National Climate Change Response Strategy (NCCRS), there is a need to integrate climate change considerations into all government planning, budgeting, and development processes at the national and county government levels. It is therefore important to formulate a range of policy instruments to address climate change. This was successfully undertaken, with a resultant national climate change action plan 2013-2017, currently being implemented (GOK, 2013). The need now is to implement the policy and to mainstream the guidelines on environmental research and development by relevant sectors, which is clarified in NEMA 2010, but not clarified in Kenya's main national research and innovation center. There is also a need to raise community awareness about CFP, clarify the CFP role in containing the societal contribution to climate change, and advocate for CFP declaration by the various companies in the country, as well as enact legislation on CFPs.

6.5 Occupational Safety and Health Services

6.51 Legislative Aspects of Occupational Health

Occupational health and safety in Kenya, including in the agricultural sector, is managed via the Occupational Health and Safety Act (OSHA), 2007. The Directorate of Occupational Safety and Health Services is the designated national authority for collection and maintenance of a database, and for the analysis and investigation of occupational accidents, diseases, and dangerous occurrences. The Directorate's policy and legal mandate are provided by the National Occupational Safety and Health Policy of 2012, OSHA 2007, and the Work Injury Benefits Act 2007, the two pieces of legislation that govern OSH services in Kenya; they are described below. The body responsible for reviewing national OSH legislation, policies, and actions is the National Council for Occupational Safety and Health (NACOSH), comprised of the Federation of Kenya Employers (FKE) and the Central Organization of Trade Unions (Kenya) (COTU-K).

6.52 The Legislative Situation of OSH

A number of legal frameworks exist which cover various mandates: As stated above, the body responsible for reviewing national OSH legislation, policies, and actions is the National Council for Occupational Safety and Health (NACOSH). The Directorate's (DOSHS) main mandate is to ensure compliance with the following:

i.   **The Occupational Safety and Health Act (2007) (OSHA 2007),** which seeks to promote safety and health at the workplace. The purpose of OSHA 2007 is to secure the safety, health, and welfare of people at work, and to protect those not at work from risks to their safety and health arising from, or in connection with, **The Act Work Injury Benefits Act, 2007 (WIBA, 2007)**, which ensures prompt compensation against work-related injuries. The purpose of WIBA 2007 is to provide compensation to employees for work-related injuries and diseases contracted in the course of their employment and for connected purposes.

ii.   **The National Occupational Safety and Health Policy of 2012**, providing the Directorate's policy and legal mandate, along with WIBA (2007) and OSHA (2007).

iii.   **The Pest Control Products Board (PCPB),** a regulatory body established under the Pest Control Products Act, Cap. 346, with a broad mandate to regulate the importation, manufacture, distribution, use, and disposal of pest control products. Its activities include registration, inspection of pesticides, and public education. In addition, the Kenya Plant Health Inspectorate Service (KEPHIS) has a mandate to protect Kenya's agriculture from pests and diseases that could have a negative impact upon the environment, economy, and human health. As provided under the Plant Protection Act (Cap. 324), all travelers are required to declare any plants, plant products, or other regulated articles carried as part of their baggage (both hand-carried products and checked-in baggage).

iv.   The Environmental Management and Coordination (Noise and Excessive Vibration Pollution) (Control) Regulations, 2009, making progress in addressing the issue of the noise as it relates to occupational health (NEMA, 2009b).

6.53 Policy Gaps in Occupational Health and Safety Services

Many OSH policies have been developed, as the above section illustrates. However, implementation of these policies is a critical challenge due to the lack of basic capacity and facilities. Basic requirements -- for example, noise monitoring, and air quality and pollutant exposure measurements at outdoor and indoor environments -- are completely absent. It is therefore not possible to associate most health conditions with any pollutant, even though pollution and exposure are clearly indicated.

6.54 Policy Needs for Occupational Health and Safety Services

There is need to collect empirical evidence linking pollution with health challenges within various segments of the Kenyan society, in both indoor and outdoor settings. This would mean gathering baseline status information on health and air pollution (including noise levels), and moving ahead to monitor the same continuously using standard procedures and equipment.

6.6: Air Pollution and Standards on Greenhouse Gas

6.61 The Environmental Management and Coordination (Air Quality) Regulations, 2008.

Working closely with KEBS, NEMA has referenced a total of 91 Kenya Standards in the draft air quality regulation, currently under publication and subsequent enforcement by NEMA. This regulation is referred to as "The Environmental Management and Coordination (Air Quality) Regulations, 2008." Its objective is to provide for prevention, control, and abatement of air pollution to ensure clean and healthy ambient air. It also provides for the establishment of emission standards for various sources such as mobile sources (e.g., motor vehicles) and stationary sources (e.g., industries), as outlined in the EMCA, 1999, and sets emission limits for various areas and facilities. It covers any other air pollution source as may be determined by the Minister in consultation with the Authority. The regulations provide the procedure for designating controlled areas and the objectives of air quality management plans for these areas.

The regulations propose that the emissions shall be controlled using specified equipment. These are air pollution control systems which are available locally and internationally from dealers. Cases of malfunctioning air pollution control systems should be reported to NEMA within 24 hours for NEMA to warn the public. Corrective measures should be taken to NEMA's satisfaction within 14 days after the occurrence. The regulations also define the methods of testing for vehicular emissions and the inspection period for motor vehicles. Private cars will be inspected every two years, and public service and commercial vehicles will be inspected annually. The Motor Vehicle Inspection Unit will identify private vehicular emission testing workshops. Standards are also set to monitor and measure air quality and carbon footprint.

6.62: ISO Global Benchmark and Air Pollution

ISO has already developed standards with an impact on climate change for areas such as building environment design, energy efficiency of buildings and sustainability in building construction, intelligent transport systems, solar energy, wind turbines, nuclear energy, and hydrogen technologies. Such GHG initiatives rely on quantifying, monitoring, reporting, and verifying GHG emissions and/or removals. With a view to implementing the task of monitoring and measuring of air quality and carbon footprint, KEBS has published a number of standards for Kenya covering the environmental sector. ISO 14064 is emerging as the global benchmark on which to base programs for reducing GHG emissions and emissions trading programs. ISO 14064 seeks to: (i) enhance the credibility, consistency, and transparency of GHG quantification, monitoring, and reporting, including GHG project emission reductions and removal enhancements; and (ii) facilitate the development and implementation of an organization's GHG management strategies and plans. Users of ISO 14064 could find benefit from (i) corporate risk management: for example, the identification and management of risks and opportunities; (ii) voluntary initiatives: for example, participation in voluntary GHG registry or reporting initiatives; (iii) GHG markets: for example, the buying and selling of GHG allowances or credits; and (iv) regulatory/ government reporting: for example, credit for early action, negotiated agreements, or national reporting programs.

ISO 14064-1 details principles and requirements for designing, developing, managing, and reporting organization- or company-level GHG inventories. It includes requirements for determining GHG emission boundaries, quantifying an organization's GHG emissions and removals, and identifying specific company actions or activities aimed at improving GHG management. It also includes requirements and guidance on inventory quality management, reporting, internal auditing, and identifying the organization's responsibilities in verification activities. The standard **ISO 14064-2** focuses on GHG projects or project-based activities specifically designed to reduce GHG emissions or increase GHG removals. It includes principles and requirements for determining project baseline scenarios and for monitoring, quantifying, and reporting project performance relative to the baseline scenario and provides the basis for GHG projects to be validated and verified.

6.63 Key Policies on Climate Change

6.631 Relevant National Policies to and Legislation regarding Climate Change in Kenya

(i) **Integrated National Transport Policy (2010):** provides for transport solutions that have relevance to climate change mitigation;

(ii) **The National Policy for the Sustainable Development of Northern Kenya and other Arid Lands** (as described above);

(iii) **The National Disaster Management Policy, 2012,** whose details are as shown above;

(iv) **Environmental Management and Coordination Act (EMCA, 1999)**: the principle instrument of Government for the management of the environment (as described above);

(v) **Water Act, 2002**: as described above;

(vi) **The Energy Policy and Act**: Kenya's energy policy of 2004 encourages implementation of indigenous renewable energy sources to enhance the country's electricity supply capacity. The policy is implemented through the Energy Act of 2006, which provides for mitigation of climate change through energy efficiency and promotion of renewable energy. In addition, the Feed in Tariffs (FiTs) policy of 2008 (revised 2012) promotes generation of electricity from renewable sources. It applies to geothermal, wind, small hydro, solar, and biomass sources of energy.

(vii) **The Kenya Forestry Master Plan 1995-2020** provides for an overarching framework for forestry development in the country for the 25 year period up to 2020, as described above.

(viii) **The Second National Environment Action Plan (NEAP, 2009-2013)** provides for a broad framework for the coordination of environmental activities by the private sector and government to guide the course of development activities, with a view to integrating environment and development for better management of resources.

(ix) **Threshold 21 (T21) Kenya** is a dynamic simulation tool designed to support comprehensive, integrated long-term national development planning. The T21-Kenya model was developed to integrate the analysis of the risks and impacts of climate change across the major sectors of the economy, society, and environment, to inform coherent national development policies that encourage sustainable development, poverty eradication, and increased well-being of vulnerable groups, especially women and children, within the context of Vision 2030.

At the international level, Kenya is party to a wide range of global and regional Multilateral Environmental Agreements (MEAs) such as:

(x) Convention for the Protection of the Ozone Layer (Vienna Convention);

(xi) United Nations Framework Convention on Climate Change (UNFCCC)-Kyoto Protocol;

(xii) Convention on Biological Diversity

6.64 Situation on Standards on Climate Change

Manifestations of climate change at both global and regional levels include the reduction in mountain glaciers, frequent and prolonged droughts, heat waves, flooding, landslides, resurgence of pests and diseases, and loss of biodiversity. ISO standards directly related to climate change include ISO 14067, now under development, for measuring the carbon footprint of products (CFP). It will complement the already published standards ISO 14064 and ISO 14065, which provide an internationally agreed framework for measuring GHG emissions, verifying claims made about them, and accrediting the bodies which carry out such activities. KEBS has published standards on energy-saving lamps covering performance and safety. This includes KS IEC 60969 Kenya standard for the performance of self-ballasted lamps for general lighting services. As stated, this is because renewable energy and energy efficient technologies offer the benefit of not only mitigating climate change but also lowering energy costs. Increasing energy efficiency of appliances and equipment in residential, commercial, and industrial sectors, for example, will reduce energy related carbon dioxide (CO_2) emissions.

The Clean Development Mechanism (CDM) is an agreed-upon framework under the Kyoto Protocol of the United Nations Framework Convention on Climate Change to assist developing countries to achieve sustainable development. Furthermore, industrialized countries under the Framework are required to comply with their GHG emissions reduction commitments. As such, projects targeted under the CDM are those activities that focus on reducing GHG emissions and also on those that enhance carbon storage (sequestration) implemented in developing countries. Kenya currently has one registered CDM project: the Mumias Sugar Company's 35MW Bagasse Based Cogeneration Project. This company is now generating 35MW of power for internal consumption, with the surplus being sold to the national grid (NEMA, 2012).

6.7 Climate change policy and enforcement -- Gaps

The GHG measurement protocol and the national energy standards for renewable energy, with regard to energy saving lamps are in place but not implemented. Relevant agencies such as NEMA need to follow up with enforcement. But due to a myriad of issues, NEMA has not been able to prepare a proper implementation framework, which is part of the reason why indoor air pollution is still a challenge for households. KEBS is mandated to develop standards; however, various agencies are responsible for implementing those standards. The OSH-related standards are operationalized by DOHSS; the environmental standards are enforced by the NEMA, and so on. The relevant agencies have met some resistance in implementing the various policies and laws, such as EMCA (1999) (NEMA, 2012). This is likely to have arisen from lack of sufficient monitoring capacity, as well as from lack of equipment and personnel.

6.8 Climate change policy and enforcement -- Needs

There is a clear need for identification of feasible options for implementing the GHG measurement for outdoor air pollution and health (including a clean development mechanism), coupled with implementing energy-saving policies for households. In addition, there is need to identify the various KEBS developed standards relevant to GEOHealth, the progress of implementation, the baseline status of each indicator, monitoring capacity, equipment, and personnel endowment of the various responsible agencies. This is known as the capacity assessment of each agency to cope with each standard it is tasked with implementing and enforcing. The capacity gaps identified in section 7.8 would be filled through multiple approaches, including: capacity building by training, seminars/workshops, acquisition of basic equipment to ensure that monitoring is feasible, and training personnel to utilize and maintain the equipment.

6.9 Health sector Strategies on Health and Climate Change

6.91 The situation

The World Health Organization (WHO) has an active and long-standing program for protecting human health from climate change, guided by a World Health Assembly resolution (WHA61.19). This resolution urges members to take decisive action to address health impacts from climate change by issuing warning of its potential risks to human health. The resolution calls on the health sector to take action to adapt projects to limit the impacts of climate change on health. The Libreville Declaration was adopted during the First Inter-Ministerial Conference on Health and Environment in Africa, held in August 2008; it represented an important step towards intersectoral actions at national and regional levels in support of the betterment of human health and ecosystem integrity on the African continent. In this Declaration, Kenya committed to implementing 11 priority action points in order to address health and environmental inter-linkages. To translate the subsequent Luanda Commitment to action, countries undertook a situational analysis and needs assessment as the basis for the development of National Plans of Joint Action (NPJAs). Intersectoral projects were selected by the Kenya Country Task Team within the Ministries of Health and Environment. These projects were:

- The Nairobi Rivers Rehabilitation and Restoration Program (NRBP);
- GOK-UNICEF Water Sanitation and Hygiene (WASH Program);
- GOK/WHO/UNDP-GEF project on climate change adaptation to protect human health in Kenya;
- Kenya National Plans of Joint Action (NPJA).

The former Ministry of Public Health and Sanitation and the Ministry of Environment and Mineral Resources were actively engaged in areas of joint action to mitigate the effects of climate change on health. This resulted in the formation of the Kenya National Plans of Joint Action (NPJA), with the objective of actualizing the commitment of the Kenya Government and all stakeholders in the improvement of the health and environment sectors. The joint activities included, among others: (i) creation of a national Health and Environment Strategic Alliance (HESA) to coordinate operations between the health and environment sectors; and (ii) review of regulatory and policy frameworks to provide coherent methods to addressing the impact of the environment on health.

6.92 Health and Climate Change Research: Gaps
However, coordination of the activities of HESA, following the elections of 2013 (which saw a reconstitution of the government with no such ministry), brought confusion in terms of institutional mandates and operations. In addition, as county governments were establishing their priorities, it was likely that a number of already established county-based mechanisms were being put in place to handle this mandate.

6.93 Health and Climate Change Research: Needs
There is a need to determine if the current agencies in national and county governments handling this mandate conduct capacity assessments and reinforce the technical and other capacities if the need arises.

7.0. KEY PRIORITY AREAS FOR RESEARCH, POLICY, AND TRAINING GAPS IN GEOHEALTH THEMATIC AREAS OF AIR POLLUTION, CLIMATE CHANGE, AND OCCUPATIONAL HEALTH

7.1 Introduction

From the previous sections on research, training and capacity, and policy, it is clear that the GEOHealth project can help to reduce the gaps in health issues related to climate change, indoor and outdoor air pollution, and occupational health and safety. The sections below summarize some of these previously-stated gaps and needs.

7.2 General Research Gaps

There is a general lack of adequate research to inform policy on higher education in climate change, occupational safety, and air pollution, with particular emphasis on their links to health. These gaps are largely attributed to lack of research opportunities, lack of collaboration among potential research partners, and lack of resources, facilities, and equipment to conduct the research. Overwhelmingly, the development of higher education is characterized by lack of access to appropriate empirical research facilities and equipment. The bulk of the research is based on theses and dissertations at the graduate level, most of which are self-sponsored and thereby characterized by major limitations arising from resource and time constraints. In addition, there is a general lack of accessible and reliable research banks and databases to inform literature reviews for new research and as a safeguard to avoid duplication. With regard to climate research, there is no database for basic meteorological data to enable the country to forecast, and therefore plan accordingly, for the impacts of climate change. In addition, the indoor and outdoor air pollution levels remain largely unknown for most settings; the existing data are sporadic and do not inform decision-making and policy, with the result that there are no reliable national data on the impacts of indoor air pollution on health in Kenya.

7.3 Current Gaps in Capacity in Climate Change Research

There is inadequate capacity in Kenya to develop nationally appropriate mitigation actions (NAMAs) and to develop and implement a sustainable GHG national inventory system. With regard to NAMAs, the challenges include lack of/or inadequate capacity to identify concrete mitigation opportunities; scenario modeling, i.e., determining future GHG emission trends under the baseline scenario (without implementation of proposed mitigation activities) and under the mitigation scenario; in evaluating external financing and support needs, as well as in implementing, monitoring, reporting, and verifying the results. The main impediments to developing a sustainable GHG inventory system are lack of relevant coordinated mechanisms between some government agencies on data sharing; institutional memory loss with dissipation of national inventory experts; lack of documenting, archiving, and reporting; and lack of quality assurance and quality control (QA/QC) and improvement strategies.

A technology registry to match the technology needs of the proposed NAMAs based on the first Technology Needs Assessment-TNA (2005) was not created to provide long-term climate change technology scenarios, as was envisaged. Kenyans are currently consumers of imported climate change mitigation technologies, e.g., Renewable Energy Technologies (RET).

Research needs and priority research areas on climate change include the following:

- Better understanding and coping with Kenya's existing climate variability;
- Better planning for future climate variability by producing climate model simulations under a range of possible greenhouse gas emission scenarios;
- Proper mitigation measures through action research in alternative energy sources in the continent, especially wind, solar power, etc., to address the likely median temperature increase for Africa which in Kenya, is estimated at 3–4°C by the end of the 21st century, or roughly 1.5 times the global mean response.

7.4 Current Gaps and Needs in Climate Change Training

Analyses of the curricula for primary, secondary, and tertiary institutions prepared by the Kenya Institute of Education (KIE) confirm that climate change has not yet been integrated into the formal education system. A survey of selected public and private universities, however, indicates that education on climate change is offered at the University of Nairobi and Kenyatta University. Other universities offer courses in environmental science and natural resources management, among others, into which climate change is progressively being infused.

The integration can be as follows:

- At primary schools, introduce children early to climate change adaptation practices using subjects such as science, nature, and -- where possible -- all other subjects;
- At secondary schools, introduce content on particular aspects of climate change relevant to the Kenyan situation: afforestation, clean energy alternatives, and climate- smart agriculture;
- At middle-level colleges, train technicians in needed skills such as solar energy systems development and maintenance; geothermal energy; wind power generation; and agriculture extension that emphasizes technologies for raising drought-tolerant crop varieties and livestock;
- At universities, infuse climate change into all professional courses.

The emerging areas for targeted capacity development already identified in the NCCRS Action Plan include climate change data collection, analysis and utilization of scenario modelling, climate change diplomacy, treaty implementation, evidence-based policy development and implementation, climate finance including carbon trading, vulnerability and risk assessments, approaches to building adaptive capacity, sector specializations to take advantage of low carbon development opportunities, and applying, monitoring, and evaluating mitigation and adaptation efforts (MRV) among others.

Some capacity for climate change MRV already exists and is mainly located within non-state actors (NSAs). The MRV/M&E of the climate change project is a new field and therefore this capacity has not yet evolved to the required levels. There is need for capacity both in terms of human resources and in terms of awareness and knowledge of climate change measures and reporting requirements and frameworks. Awareness of climate change indicators and reporting requirements is also generally low. Therefore, efforts to increase this knowledge through awareness-raising strategies and training are key to building capacities in this area.

Kenya lacks middle-level technicians to support widespread adoption of technologies that will be required to support the country's aspirations towards a low carbon development pathway. For example, while it is desirable for Kenyans to adopt solar technology on a mass scale, supportive networks for installation and maintenance of solar power systems are not well-established nationally. Furthermore, standards for such system are non-existent. The same can be said of other technologies, such as those to harness geothermal and wind energy. Therefore opportunities exist to train technicians with these skills to support the widespread adoption of adaptation and mitigation practices needed to support the NCCRS Action Plan.

At the university level, climate change should be infused into the various professions:

- Civil engineers need to learn how to design and develop structures that can withstand climate shocks;
- Physicians need to be aware of the effects of climate change on human health;
- Architects should have the skills and training to design houses that are climate-proofed, climate change resilient, and energy-efficient;
- Teachers ought to be equipped with knowledge about climate change for them to be prepared to teach a curriculum that integrates climate change across all subjects taught at schools in Kenya.

7.5 Training Gaps

Although a substantial number of local universities offer undergraduate and post-graduate degrees in climate change, air pollution, and occupational health, a number of weaknesses were identified among the graduates of the institutions, including poor-problem solving skills, poor time management, lack of innovation, lack of exposure to practical experiences, and lack of positive attitude on the part of the graduate regarding the area of specialization. The majority of these arise from lack of exposure and unnecessary emphasis on theoretical approaches during training.

7.6 Priorities in Training

More proactive training with a paradigm shift in approach is recommended for the students in these particular areas, along with more practical experience during training. Most importantly, practical training encompassing field visits, local and international industrial attachments, and internships, would make these programs more effective. These changes could be coupled with pragmatic training involving the establishment of well-equipped laboratories to study and monitor climate change, occupational health, atmospheric pollution, and indoor pollution caused by the lighting and heating systems. Training should also actively involve communities and policy makers.

7.7 Occupational Health and Safety (Agriculture)

There are no baseline occupational health data in Kenya. Currently, annual disease burden records within the formal sector indicate that the sugarcane and flower sectors pose the most agriculture-based occupational health risks. Other significantly-affected demographic groups are traffic police and informal sector laborers, such as cobblers, vehicle mechanics, hawkers, and vegetable sellers, among others, working along busy highways. In the short run, an occupational health monitoring and evaluation framework is needed, especially with focus on its relationship to levels of air pollution, including cover noise and air quality. In the long run, however, an occupational health monitoring and research laboratory facility is needed to make significant progress in this thematic area. Currently, it is a stagnated area, with the occupational burden of diseases increasing at an alarming rate.

7.8 Summarized Research and Policy Priority Needs

i.   Ensure adequate collaborative research among professors, graduate students, policy makers (e.g., those affecting occupational health workers), and other stakeholders to inform policy on higher education relevant to climate change and air pollution, and their links with health.

ii.   Focus on occupational health, especially on the flower and sugarcane industries which are the fastest growing areas where the dangers are the strongest; with regard to air pollution and health, the focus should be on roadside workers (Jua kali / hot sun/ informal sector workers, informal / small scale garage workers, traffic police, schools in the middle of the city, etc.).

iii. Focus on indoor air pollution through collaborative intervention research between community/ households and universities, NGOs, and government departments which handle housing issues. Indoor air quality could be monitored with a focus on intervention research among the households in various socio-economic and cultural settings. A long-term focus could be a housing and energy-source package which best suits the various environments, designed following a series of trials, with indoor air quality and health monitoring among the various age groups, gender, and the household members conducted over the life of the GEOHealth project.

iv. Develop empirical research capacity at the university level through partnerships with other stakeholders to conduct the above research agenda. This should have a built-in graduate research component to enable students contribute to the broader research and policy framework.

v. Establish partnerships with reputable facilities to support empirical research at all levels of university programs and other research institutions. Basic portable monitoring equipment such as the HACH air quality monitoring equipment could be used in the short run. A long-term goal, e.g., after two to three years, would be the establishment of reliable research-oriented equipment such as spectrophotometers, followed within one to two years by the establishment of a fully-fledged air pollution and occupational health laboratory with a broad range of capacities to monitor a wide range of parameters. This could include a state-of-the-art GIS laboratory or a national environmental health laboratory.

vi. Implement regular monitoring and establish accessible and reliable research banks and databases to inform new research and to avoid duplication, especially on data related to the three themes of GEOHealth. These should include a database for basic meteorological data to enable the country predict and therefore plan for climate change; a database for county / regional indoor and outdoor air pollution levels; occupational health laboratory; and national data on impacts of indoor air pollution on health in Kenya.

vii. Strengthen institutional capacity to prepare inventories and establishing a trained, sustainable inventory team to help Kenya reduce uncertainties and improve the quality of inventories for subsequent national communications.

viii. Build indigenous capacity to innovate, design (with the help of NACOSTI, KIRDI and KEBS), and add value to renewable energy and other heating, cooking, and lighting technologies devoid of indoor pollutants with impacts on health. With regard to solar products, for example, Kenya and the region should be building infrastructure to service products and adapt them to the Kenyan and regional context. This is an area where the region has a comparative advantage due to its vast untapped alternative energy sources such as sunlight. The types of institutions, policies, and stakeholder networks that will assist to this end should be identified, and the proposed GEOHealth Hub can play a significant role in bridging this clear gap and need.

8.0 CONCLUSIONS AND RECOMMENDATIONS

8.1 Introduction

Kenya, like the rest of the world, is experiencing high levels indoor and outdoor air pollution from its anthropogenic activities. These are in turn resulting in excessive impacts of climate change and variability, as well as increasing the environmental and occupational burden of disease. Climate change and variability affect natural processes which lead to an increased incidence of a range of diseases, such as asthma, malaria, diarrhea, Rift Valley Fever, and nutrition-related ailments. Climate change affects ecosystems, water resources, food, health, coastal zones, industrial activity and human settlements. Addressing these impacts, however, presents opportunities for innovation, business growth, and improved livelihoods.

8.2 Research and its Output

A number of universities offer courses and research in climate change, occupational health, and air pollution at the post-graduate level. As stated, these are mostly self-sponsored, and their impacts hardly go beyond the thesis defense and graduation. The key reason that research data from the theses are not used by policy makers is their unreliability, methodological flaws, temporal and spatial limitations occasioned by resource and facility limitations, and lack of relevance to policy needs. The policy institutions also experience financial constraints, and are unable to team with other stakeholders to conduct more relevant research.

Most of the thesis research work produces data using inappropriate or outdated equipment, rendering such data unable to provide proper evidence to inform policy. Moreover, there is no research consistency and continuity on any given subject. As such, the data are one-off and there is no means to guarantee that a research line can be maintained over a period to to establish a trend. Even if the data collected are related, they are from different geographical settings, and the research is more commonly conducted using different methods. As such, they cannot provide a solid evidence base for policy.

Beyond the research, there are further complications in linkages to policy and stakeholder consumption. In Kenya, most research studies take 3-10 years before appearing in a readily accessible, peer-reviewed journal. By the time the results are published, they are outdated and hence unable to inform reliable policies. In addition, most are not indigenous, or locally supported, and not attached to Kenyan institutions; they are therefore not sustainable. As such, the key sources of continuous research data informing local policy are sporadic research studies that are supported a year of less by non-Kenyan sources, leaving no appreciable sustainability components behind.

There is a need to coordinate research through a well- functioning center responsible for identifying research themes and focusing on a logical sequence / flow with a clear spatial and temporal trend. Such data would need to be generated on a systematic and continuous basis via proper field work and laboratories for well-guided research, with a database and fulltime analytical tools, and with means to share the data. The center should also incorporate indigenous knowledge systems which most communities in Kenya have used traditionally to escape disasters. The Maasai community which hosts the Maasai Mara University is very rich in this aspect, and has demonstrated that it can be an asset in such research.

8.3 Research-Policy Interface

Policies on indoor and outdoor air pollution, climate change, and occupational health (agriculture) and their means of enforcement are clearly inadequate in Kenya. At least 99% of the policies have international origins, with less than 1% resulting from local research at academic institutions. A majority of this 1% is sponsored by international institutions and partners. As this is the route that has worked for Kenya, the developing network should capitalize on this method and its impact by engaging with all sectors of stakeholders to collectively conduct empirical and other research and policy advocacy in the three thematic areas. This process could be used to train students at all levels so that the foundation is strong enough for them to engage at whatever level they choose, including research, upon graduating and embarking on full-time work. This would mean conducting a number of workshops with wide sectoral representation to formulate a research and policy framework and a plan on how training on the three thematic areas can be a tool to make the process work. Lastly, once the research is done, its results need to be communicated through various mechanisms such as the electronic and print media, conferences, workshops, and policy briefs. The GEOHealth Hub should attempt to establish such a system to promote and simplify the communicating of research data.

8.4 Climate Change, Greenhouse Gas Emissions, and Air Quality

There are no reliable data on climate change in Kenya. The basic meteorological data are unreliable, as well as unavailable to would-be users. This barrier makes it difficult for stakeholders such as households, capacity-building learners (such as universities), and policies makers (including government) to make decisions. As such, there is an urgent need to generate and store clear empirical, research data in an easily retrievable information system.

Based on the above scenario, there is a substantial lack of policies in the areas of air pollution and climate change, and stakeholders urgently need to formulate policies to address these issues. However, the Kenya Bureau of Standards (KEBS) has attempted to deal with this absence by implementing a number of standards, such as ISO 14064, which is emerging as the global benchmark on which to base programs for reducing GHG emissions and emissions-trading programs.

8.5 Institutional Linkages, Research, and Policy

As most of the research and policy institutions do not have strong links with universities, especially in the areas of climate change and air pollution, as well as occupational health and safety, there is need to develop frameworks and strategies to facilitate better linkages and partnerships between research institutions, universities, policy institutions (largely encompassed by government parastatals and departments,) and other stakeholders such as community groups (CBOs) and civil society. It would be helpful to bring together the various institutions to collaborate on a joint and harmonized linkages platform and policy to establish some degree of uniformity and to minimize future conflict.

8.6 Training and Research

As stated earlier, a significant number of the local universities offer training and research degrees in climate change and in occupational health, but their weaknesses need to be addressed, especially the lack of exposure to practical training and an overly theoretical approach to studies. As such, more proactive and pragmatic training, including the active involvement of communities and policy makers, is recommended and should include local and international attachments and internships, coupled with the establishment of equipped laboratories to study and monitor climate change, occupational health, atmospheric pollution and indoor pollution caused by the lighting and heating systems. Opportunity exists with the younger, fast-growing institutions of higher learning as there is scope for developing a more sustainable foundation devoid of bureaucratic red tape which can bog down projects as often happens in the older institutions, where problems remain unidentified and minimal effective impact is made on the three thematic areas. However, the young institutions, as much they have more growth prospects, also still have serious challenges of lack of strong institutional frameworks to manage some types of research. But where they face the challenges with an open mind, they can achieve much more. In Kenya, this is where Maasai Mara University, University of Kabianga etc.

Comparing the two, for instance, Kabianga has a decentralized management system, with various schools, including the school of science where there are health programs (including environmental Health) based in Kapkatet campus. It also has the school of environment where the climate change issues are handled. In addition, it has clearly marked research as a key priority and appointed a director of research with an office. This makes research issues easier to manage and coordinate. Maasai Mara University, on the other hand, has all its programs in one campus, with a lot more centralized management, but no health program yet. It has no research office, with everything coordinated from the deputy vice chancellor's office, which is rather busy at the same time with student affairs and academic affairs (where day to day management of classes and academic programs are managed from) which make the office too heavy for the three core areas of jurisdiction. Research issues therefore still lie at the back. A SWOT therefore reveals that even the universities of the same age are at different levels depending on their prioritization of various responsibilities.

8.7 Air Pollution and Health

In Kenya, very little reliable empirical research has been carried out in this field, rendering most available data scattered and unreliable. Most of what is available is too sporadic to inform stable policies. There is no consistent indoor and outdoor air quality monitoring system and this gap renders associating such pollution with health almost impossible. As such, there is need to establish indoor and outdoor air quality monitoring systems, coupled with baseline health status of those who work in those specific environments, e.g., traffic police for outdoor air pollution and health, and women and children in various parts of Kenya for indoor air pollution and health. This health monitoring system should include an intervention package in which the impacts of differences among treatment levels (e.g., with eaves spaces), fuel types, house sizes, time and duration of stay in house/ street, work shifts, meal times and numbers, where meals are taken, and family size are evaluated. These associations can be established after a consistent period of data collection and can be carried out through collaboration among professors, graduate/research students, policy makers, and the primary stakeholders. This would mean some impacts and the appropriateness of various interventions in different socio-economic settings should be evaluated, and these interventions could include, but not be limited to, use of solar/ wind energy and variants of cook stoves. Some local capacity building towards adopting and encouraging communities to use technologies confirmed to promote a healthier environment may need to be incorporated into the research program.

8.8 Occupational Health

Occupational health in Kenya is largely addressed through the Occupational Safety and Health Act (OSH Act 2007), by the Directorate of Occupational Safety and Health Services (DOSHS). However, because of socio-economic constraints, Kenyan citizens seek jobs where they can be found, rarely considering occupational health threats. As such, there is need for continuous monitoring of the health status of workers, especially the health effects that relate to their particular type of work, and the specific assignment they are employed to do. This should be coupled with regular health monitoring, recording, analysis, and providing feedback to the institution and/or the employee with the position of taking more proactive means of protecting themselves from work-related hazards. The need exists to formulate specific policies on occupational health and safety for the agricultural and informal sectors, which are the most vulnerable with regard to the occupational burden of diseases.

It is clear that the bulk of injuries and long-term health impacts originate from the agricultural sector and that the occupational health of workers in this sector must be enhanced through intervention research. This could include health monitoring, beginning with data collection of baseline health status. The focus here, and for subsequent policy, could be in the sugarcane and flower sectors, as their workers are currently the most exposed to occupational health challenges. Such interventions and subsequent policies could have positive effects on the carbon footprints of related products, and could result in institutionalizing the reduction of pollution by private-sector partners.

9.0. REFERENCES

1.   Bates, Elizabeth (2002): Smoke health and household energy Issues paper compiled for DFID – EngKaR project no. R8021– September 2002. ITDG 2002.

2.   Bates, Elizabeth, Nigel Bruce, Alison Doig, Stephen Gitonga. Participatory approaches for alleviating indoor air pollution in rural Kenyan kitchens. 2002. ITDG, The Schumacher Center for Technology & Development.

3.   Bailis R, Ezzati M, Kammen DM. 2005. Mortality and greenhouse gas impacts of biomass and petroleum energy futures in Africa. Science 308(5718):98-103.

4.   Boiling Point No 48 2002 13: Boiling Point 48/final 6/11/02 1:52 pm Pg 13.

5.   Central Organization of Trade Unions (Kenya) (COTU-K): http://www.cotu-kenya.org

6.   Clark ML, Peel JL, Balakrishnan K, Breysse PN, Chillrud SN, Naeher LP, Rodes CE, Vette AF, Balbus JM. 2013. Health and household air pollution from solid fuel use: the need for improved exposure assessment. Environ Health Perspect 121:1120–1128; http://dx.doi. org/10.1289/ehp.120642

7.   Directorate of Occupational Health and Safety (DOSHS) Annual and Strategic Reports.http://www.doshs.go.ke

8.   Egondi, Thaddaeus, Catherine Kyobutungi, Nawi Ng, Kanyiva Muindi, Samwel Oti, Stephen van de Vijver, Remare Ettarh and Joacim Rocklov (2013). Community perceptions of Air pollution and related Health risks in Nairobi Slums. Int J Environ Res Public Health. 2013 Oct; 10(10): 4851–4868. Published online 2013 Oct 11. doi: 10.3390/ijerph10104851.

9.   EPA: Vehicle technologies, particularly the use of catalytic converters. EPA: Africa Air Quality: http://www.epa.gov/international/air/africa.htm

10.   EPA: Promoting Cleaner Fuels and Vehicles Worldwide: Lim, et. al., Global Burden of Disease Study 2010, Lancet, Vol 380 December 15/22/29.

11.   Ezzati, M., D. M. Kammen, and B. H. Singer (1999) "The Health Impacts of Exposure to Indoor Air Pollution from Biofuel Stoves in Rural Kenya," The Proceedings of Indoor Air 99: the 8th International Conference on Indoor Air Quality and Climate; Edinburgh, Scotland; August 1999, 3, 130-135.

12.   Ezzati, M., Mbinda, B.M., and Kammen, D.M. (2000) "Comparison of Emissions and Residential Exposure from Traditional and Improved Cookstoves in Kenya," Environmental Science and Technology (ES&T), 34 (2), p. 578-583.

13.   Ezzati, M., Saleh, H., and Kammen, D. M. (2000) "The Critical Role of Microenvironments in Pollutant Exposure and Impact: Individual Behaviour and Indoor Air Pollution from Biomass Combustion in Kenya," Environmental Health Perspectives, submitted.

14.   Ezzati M, Lopez AD, Rodgers A, Murray CJL. Comparative quantification of health risks: global and regional burden of disease attributable to selected major risk factors. Geneva, World Health Organization, 2004.

15.   Galcano C. Mulaku and L. W. Kariuki: Mapping and Analysis of Air Pollution in Nairobi, Kenya. International Conference on Spatial Information for Sustainable Development Nairobi, Kenya 2–5 October 2001.

16.   Government of Kenya/Ministry of Environment and Natural Resources. 2010. National Climate Change Response Strategy. Nairobi.

17.   Government of Kenya/ Ministry of Environment (GOK, 2014). National climate change action plan 2013-2017.

18.   ITDG and UK Department for International Development (DFID), ITDG: Reducing indoor air pollution in rural households in Kenya: working with communities to find solutions. The ITDG Smoke and Health project 1998-2001.

19.   Kammen, DM,. Ezzati M, Mbinda B.M. 1999. The Determinants of Exposure to Indoor Air Pollution from Biofuel Stoves in Rural Kenya, The Proceedings of Indoor Air 99: the 8th International Conference on Indoor Air Quality and Climate; Edinburgh, Scotland; August 1999, 3, 171-176.

20.   Kammen R, Ezzati M: indoor air pollution from biomass combustion and ARI in Kenya: an exposure response study, Washington DC, USA, Center for Risk Management 2001- PubMed.

21.   KDHS 2006: Disease prevalence in Kenya 2010.

22.   Kenya Climate Change Action Plan (2012).

23.   Kenya Institute for Public Policy Research and Analysis. 2010. A Comprehensive Study and Analysis on Energy Consumption Patterns in Kenya: A Synopsis of the Draft Final Report. Nairobi: Kenya Institute for Public Policy Research and Analysis.

24.   Kenya National Cancer Control Strategy: Kenya 2011-2016.

25.   Kinney, Patrick L., Michael Gatari Gichuru, [...], and Elliott Sclar (2011) Traffic Impacts on PM2.5 Air Quality in Nairobi, Kenya. Environ Sci Policy. June 2011; 14(4):369-378.

26.   Kipkorir BS (2013): Determination of Selected Indoor and Outdoor Air Pollutants in the Central Business District of Nairobi City, Kenya. A Thesis Submitted In Partial Fulfilment For The Requirements Of The Degree Of Master Of Science In Applied Analytical Chemistry In The School Of Pure And Applied Sciences Of Kenyatta University; February 2013.

27.   Magadza, C.H.D. 2000. Climate Change Impacts and Human Settlements in Africa: Prospects for Adaptation. Environmental Monitoring and Assessment, 61(1), 193-205.

28.   Martin WJ II, Glass RI, Araj H, Balbus J, Collins FS, et al. (2013) Household Air Pollution in Low- and Middle-Income Countries: Health Risks and Research Priorities. PLoS Med 10(6): e1001455. doi:10.1371/journal.pmed.10014559.

29.   Moturi N W (2010). Risk factors for indoor air pollution in rural households in Mauche division, Molo District, Kenya. Afr Health Sci.: Sept 2010 10(3): 230-234. accessed from: http://www.ncbi.nlm.nih.gov/pmc/articles/PMC3035965/

30.   National Climate Change Response Strategy Stakeholders' Workshop Reports. 2009.

31.   National Occupational Safety and Health Policy 2012.

32.   NEMA. 2008. Effects of climate change and coping mechanisms. State of the Environment report Kenya 2006/7.

33.   NEMA (2012) NEMA news: a quarterly publication. The National Environment Management Authority, Nairobi. NEMA (2009). Draft Air quality regulations, 2009. Nairobi, Kenya.

34.   NEMA (2009b). The Environmental Management and Coordination (Noise and Excessive Vibration Pollution Control) Regulations, 2009.

35.   Occupational Safety and Health Act 2007.

36.   Odhiambo, G.O., A.M. Kinyua, C.K. Gatebe and J. Awange (2010). Motor Vehicles Air Pollution in Nairobi, Kenya. Research Journal of Environmental and Earth Sciences (Res. J. Environ. Earth Sci., 2(4): 178-187, 2010 ISSN: 2041-0492 © Maxwell Scientific Organization, 2010 Published Date: October 05, 2010.

Patz, J. (2000). The potential impacts of climate variability and change for the United States: Executive summary of the report of health sector of the National Assessment. Environmental Health Perspectives, 108(2000), 367-376.

37.   Responses for Kenya. Poverty Reduction and Economic Management Unit Africa. Nairobi.

38.   Saleh, H., M. Ezzati, D. M. Kammen, and B.H. Singer (1999) "The Health Impacts of Exposure to Indoor Air Pollution from Biofuel Stoves in Rural Kenya: A Behaviour Based Analysis," Submitted to Environmental Science and Policy.

SEI (2009). The Economics of Climate Change in Kenya. Stockholm Environment Institute. Oxford.

39.   Sessional Paper on Integrated National Transport Policy; Ministry of Transport, November 2010.

40.   United Nations Population Division. World Population Prospects - the 2006 revision.   New York: United Nations, 2007.

UNEP (2009). Kenya: Atlas of Our Changing Environment. Division of Early Warning and Assessment (DEWA), United Nations Environment Programme (UNEP). Nairobi, Kenya.

41.   Warau (2014) http://www.nation.co.ke/oped/Opinion/Africa-cities-burden-of-cancer-causing-air-pollution--/-/440808/2264326/-/8mmkxi/-/index.html. Monday, March 31, 2014: FILE NATION MEDIA GROUP.

42.   Weadapt:http://weadapt.org/knowledge-base/national-adaptation-planning/ kenya

43.   Work Injury Benefits Act 2007. Republic of Kenya.

44.   World Bank. 2011: The Drought and Food Crisis in The Horn of Africa: Impacts and Proposed Policy.

45.   World Health Organization (2005): Household air pollution and health: WHO Fact sheet N°292: Updated March 2014. http://www.rff.org/Publications/WPC/Pages/09_15_08%20Indoor%20Air%20Pollution%20and%20Africa%20Death%20Rates.aspx. Accessed 15 June 2015.

46.   World Health Organization (2013). International Agency on Cancer Press release No 221, Lyon and Geneva. 17th October, 2013.

47.   World Health Organization (2009) Global Health Risks Summary Tables, October 2009. Health Statistics and Informatics Department, World Health Organization, Geneva, Switzerland. http://www.who.int/evidence/bod

48.   World Health Organization (2009). Global health risks: mortality and burden of disease attributable to selected major risks.   Geneva, World Health Organization, 2009. Available at http://www.who.int/evidence/bod

49.   World Health Organization (2008). The global burden of disease: 2004 update.   Geneva, World Health Organization, 2008. Available at http://www.who.int/evidence/bod

50. Y. von Schirnding, N. Bruce, K. Smith, G. Ballard-Tremeer M. Ezzati, K. Lvovsk (2002).

Addressing the Impact of Household Energy and Indoor Air Pollution on the Health of the Poor. Implications for Policy Action and Intervention Measures. Paper Prepared for the Commission on Macroeconomics and Health. (Based in parts on the proceedings of a WHO-USAID Global Consultation on the Health Impact of Indoor Air Pollution and Household Energy in Developing Countries, Washington DC, 3-4 May 2000, Working group 5: Improving Health Outcomes of the Poor). WHO/HDE/HID/02.9.

10.0 APPENDICES

Appendix 1: Table 0: Examples of institutions offering some courses related to the three themes of GEOHEALTH

University/ institution	BSc				MSc				PhD			
	Climate	Air pollution	OSH	Policy	Climate	Air pollution	OSH	Policy	Climate	Air pollution	OSH	Policy
KU	Conservation Biology; Env Planning & Mx env Science. Env Science: Env studies (Resource conservation. community development.	Analytical chemistry	Occupational safety and health; Env Health.		Geography (climatology); Env Planning & Mx env Science. Env Science: Env studies (Resource conservation. community development: env education/climate change and sustainability); renewable energy technology		MPH; occupational safety and health	Geography (urban and regional planning)	Geography (climatology); Environmental Physics applied analytical chemistry. Env Planning & Mx. env Science. Env Science Env studies (Resource conservation. community development: env education/climate change and sustainability); renewable energy technology			Geography (urban and regional planning); Public health; occupational safety and health; renewable energy technology
Eldoret	Env Science (Science and Art; Sustainable energy and climate change; natural resources	Chemistry			Information systems: Forestry	Biology and Health	Biology and Health Health	Planning and management				
JOOUST	Renewable energy and technology; Geography and natural resources management:			Development and policy studies				Urban env planning & management; Water and env management:	Urban env planning & management; Water and env management:			Urban env planning & management; Water and env management:
Karatina	Forestry; natural resources env science				Env Biology; Env Health:; forestry	Env Biology; Env Health:	Env Biology; Env Health:	env Economics & and Env planning	Env Biology; Env Health:	Env Biology; Env Health:	Env Biology; Env Health:	env Economics; & and Env planning

78

Appendix 2: Table 1: Research sampling frame

region	Institutions	Sector	Theme
Rift Valley	Universities 1. Egerton 2. Moi 3. Eldoret 4. Maasai Mara (RCE) 5. University of Kabianga 6. Kabarak	1.Education / training 2.Research 3. Advocacy	1. Indoor air pollution 2. Outdoor air pollution 3. Climate change 4. Occupational health and safety (agricultural)
Western and Nyanza	a. Universities 1. Maseno 2. Masinde Muliro 3. GLUK b. EAC / IUCEA c. KEMRI	1.Education / training 2.Research 1. Policy 2. Research	1. Indoor air pollution 2. Outdoor air pollution 3. Climate change 4. Occupational health and safety (agricultural)
Nairobi and Central Kenya	Universities 1. Nairobi (Chiromo campus, Medical school, school of public health, Kabete campus (CAVS)) 2. Kenyatta (School of Environ, school of Public Health) 3. Jomo Kenyatta (JKUAT) 4. Kenya Methodist University (KEMU) 5. Mount Kenya University (MKU) 6. Baraton 7. Kereri Women's University/ Gretsa 8. Strathmore 9. USIU	1.Education / training 2.Research	1. Indoor air pollution 2. Outdoor air pollution 3. Climate change 4. Occupational health and safety (agricultural)

		1. Policy 2. Research 3. Advocacy	1. Indoor air pollution 2. Outdoor air pollution 3. Climate change 4. Occupational health and safety (agricultural)
Nairobi	a. Policy/ Research institutions 1. IPPRA 2. Ministry of Health (Research Div) 3. NEMA 4. KEBS 5. DOHSS 6. CUE 7. NACOST 8. AMREF 9. KEMRI 10. KIRDI b. Advocacy organizations 1. COTU - K 2. UASU 3. FKE 4. KNUT		

Appendix 3: Table 2: Data collection tool for higher learning institutions

Institution	Name of course offered	Level of course offered 1. BSc 2. MSc 3. PhD 4. Other (specify)	Classification of course 1. Air pollution 2. Climate change 3. Occup health 4. All	How long course has been offered	Number of graduates per year	Number of graduates so far	Where graduates employed	Outstanding graduates of program
1								

Appendix 4: Table 3: Data collection tool for Research institutions

Institution	Title of degree program	Level of Research offered 1. MSc 2. PhD 3. Other (specify)	Classification of research 1. Indoor Air pollution 2. Outdoor air pollution 3. Climate change 4. Ocup health	How long research by course has been offered	Number of research undertaken per year	Number of research done so far	Where research applied	Outstanding research innovations of program

Appendix 5 Table 4: Data collection tool for individual research items

institution	Title of research	Level of Research offered 1. MSc 2. PhD 3. Other (specify)	Classification of research 1. Indoor air pollution 2. Outdoor air pollution 3. Climate change 4. Ocup health	Where research applied	Outstanding research innovations of The research	Sponsor of research

Appendix 6: Table 5: Source of policy and link with research activities in the country

Institution	Title of Policy	Source of policy: If from research work, specify main source	Level of Research that informed the policy 1. MSc 2. PhD 3. Other (specify)	Classification of Policy 1. Indoor Air pollution 2. Outdoor air pollution 3. Climate change 4. Occup health	Where research that informed the policy was undertaken (host institution)	Sponsor of research that informed the policy

Appendix 7: Table 6: The Institutions relevant to GEOHealth and their core business

	ORGANIZATION	ESTABLISHED	NATURE	THEMATIC FOCUS	RELEVANT CORE FUNCTIONS
1	The National Environmental Management Authority (NEMA)	2002, under the Environmental Management and Coordination Act, EMCA 1999	Policy	Environment (Climate, air pollution)	• Coordinate the various environmental management activities being undertaken by the lead agencies • Promote the integration of environmental considerations into development policies, plans, programs and projects, with a view to ensuring the proper management and rational utilization of environmental resources, on sustainable yield basis, for the improvement of the quality of human life in Kenya. • Take stock of the natural resources in Kenya and their utilization and conservation. • Examine land use patterns to determine their impact on the quality and quantity of natural resources. • Carry out surveys, which will assist in the proper management and conservation of the environment. • Advise the Government on legislative and other measures for the management of the environment or the implementation of relevant international conventions, treaties and agreements. • Advise the Government on regional and international conventions, treaties and agreements to which Kenya should be a party and follow up the implementation of such agreements

#	Organization	Description	Research and data management	Climate and air pollution
2	The Kenya Meteorological Service	Department in a government Ministry; it monitors, analyses, reports and records the weather patterns.			• Provision of meteorological and climatological services to all relevant sectors; • Organization and administration of surface and upper air meteorological observations within its area of responsibility and the publication of climatological data; • Co-ordination of research in meteorology and climatology including co-operation with other authorities in all aspects of applied meteorological research, and the maintenance of the National Meteorological Library; • Evolvement of suitable training programs in all fields of meteorology and other related scientific subjects which are relevant to the development of Kenya and other countries that participate in the Department's training activities.
		Climate Change And Pollution Monitoring Services In The Kenya Meteorological Department			Climate Change and Pollution (CCP) Monitoring Services is among the new Sub-Branches in the Department, under the Forecasting and Regional Office Branch. The main objectives of this sub-branch are: a) To undertake Climate Change Monitoring, Detection and Attribution especially in terms of the trends and occurrence of severe weather and extreme climate events; b) To monitor background atmosphere and urban air pollution; c) To conduct Vulnerability assessments to Climate Change; and d) To carry out Mitigation and Adaptation options/strategies to Climate Change. to achieve the Sub-Branch mandate, the following Divisions / Sections / Units have been constituted: (i) Climate Change; (ii) Atmospheric Pollution Monitoring; (iii) Urban Pollution Monitoring.

	KMD	II. Atmospheric Pollution Monitoring a) Mt Kenya GAW Station		KMD operates the Mt. Kenya Global Atmosphere Watch (GAW) Station [3678 metres altitude, 0°3'S 37°18'E] that is part of the WMO Atmospheric Research and Environment Program (AREP, It was established in 1996 through funding by the Global Environmental Facility (GEF) of World Bank and the government of Kenya. The station is designated for long-term measurements of atmospheric chemical composition and background pollution that includes including providing vital information on greenhouse gases (GHGs) and aerosols in Equatorial Africa, and the effect of biomass burning on the regional build-up of tropospheric ozone.
3	Ministry of Health Kenya	Environment Department through the unit of Climate Change and Health.	policy	Health	• The wide ranging effects of climate change include increase in water and vector borne diseases, malnutrition and livestock starvation, and land degradation. Implements the WHO resolution on climate change.
4	The African Medical research Foundation (AMREF)			health	AMREF's main areas of intervention are maternal and child health; HIV and Tuberculosis; safe water and sanitation; malaria; and essential clinical care. Amref Health Africa shares knowledge gained from grassroots programmes with others, and uses it as evidence to advocate appropriate change in health policy and practice. In all our programmes, Amref Health Africa partners with communities, civil society organisations, health practitioners, and the private and public sectors to establish a participatory health care system.
5	The Kenya Bureau of standards (KEBS)			Standards (all)	Is the National standards body responsible for the establishment and maintenance of Kenya Standards. The National Standards Council is responsible for overseeing KEBS' strategic and policy affairs. KEBS 2007-2012 strategic plan has broad objectives that include facilitation of trade, the realization of Kenya's social and environmental priorities through standardization.

#				
6	The National Commission for science, technology and innovation (NACOSTI)	Established under the Science, Technology and Innovation Act, as successor of the National Council for Science and Technology (NCST). The Act facilitates the promotion, coordination and regulation of the progress of Science, Technology and Innovation (ST&I) in the country.	Research policy	all	• Liaise with the Kenya National Innovation Agency and the National Research Fund to ensure funding and implementation of prioritized research programs. • Ensure coordination and co-operation between the various agencies involved in science, technology and innovation research and development activities. • Assure relevance and quality of science, technology and innovation programs in research institutes. • Advise the Government on policies and any issue relating to scientific research systems. • Promote the adoption and application of scientific and technological knowledge and information necessary in attaining national development goals. • Develop and enforce codes, guidelines and regulations in accordance with the policy determined under this Act for the governance, management and maintenance of standards and quality in research systems. •
7	The Kenya institute of policy analysis (KIPRA)	KIPPRA is an autonomous public institute established in May 1997 through a Legal Notice and commenced operations in June 1999.	Research, policy	training	• Conducts objective research and analysis on public policy issues with the goal of providing advice to policy makers; collects and analyses relevant data on public policy and disseminates its research findings to a wide range of stakeholders through workshops/ conferences; undertakes contracted public policy research and analysis for the government and clients from the private sector.

#	Institution		Type	training	
7	The Kenya institute of policy analysis (KIPRA)	KIPPRA is an autonomous public institute established in May 1997 through a Legal Notice and commenced operations in June 1999.	Research, policy	training	• Conducts objective research and analysis on public policy issues with the goal of providing advice to policy makers; collects and analyses relevant data on public policy and disseminates its research findings to a wide range of stakeholders through workshops/conferences; undertakes contracted public policy research and analysis for the government and clients from the private sector.
8	The directorate of Occupational Health and safety services (DOHSS)	The Directorate's policy and legal mandate are provided by the National Occupational Safety and Health Policy of 2012, OSHA 2007, and WIBA 2007.	Policy	Occupational health, air pollution	To ensure compliance with the provisions of the Occupational Safety and Health Act, 2007 that seeks to promote safety and health at the workplace. It also ensures compliance with the provisions of the Act Work Injury Benefits Act, 2007 through prompt compensation of employers against work-related injuries. DOSHS is the designated national authority for collection and maintenance of a database, and for the analysis and investigation of occupational accidents and diseases, and dangerous occurrences.
		OCS in the Agricultural Sector: The Pest Control Products Board (PCPB)			• Is a regulatory body established under the Pest Control Products Act, Cap. 346. Its broad mandate is to regulate the importation, manufacture, distribution, use and disposal of pest control products. Its activities are outlined below: • Public education and awareness creation: The participants are educated on the proper, safe use of pesticides as a critical subject, to ensure the protection of human health against risks.
		Role in Environment and Exposure Monitoring			• It is the national body that has regulatory responsibilities in environment and exposure monitoring, medical examination, surveillance of workers' health, and advisory services. • Other agencies include the National Environment Management Authority (NEMA), the Ministry of health, and the public health departments in the local authorities. • The Occupational Health Division in DOSHS undertakes occupational health surveillance in workplaces. It also monitors and supervises the

#	Institution		Category	Details
		National Council for Occupational Safety and Health (NACOSH).		activities of the designated health practitioners who carry out medical examination of workers. • The Directorate's Occupational Hygiene and Occupational Health divisions are responsible for analytical and assessment work related to the determination of workers' exposure to various occupational hazards. The body responsible for reviewing national OSH legislation, policies and actions; Composition includes the Federation of Kenya Employers (FKE) and the Central Organization of Trade Unions (Kenya) (COTU-K).
9	The Kenya Industrial Research development institute (KIRDI)	KIRDI was established in 1979 under the Science and Technology Act Cap.250 as a multidisciplinary Institution to conduct Research and Development in Industrial and Allied Technologies.	Policy, research	• The primary functions of KIRDI are derived from the institution's mandate as stipulated in the Science and Technology Act Cap 250, Laws of Kenya Section 14. These include: • To co-operate with other organizations and institutions of higher learning in training programs on matters of relevant research. • To liaise with other research bodies within and outside Kenya carrying out similar research. • To disseminate research findings.
		The Environment Management Division is one of the research, technology and innovation (RTI) departments in KIRDI		The division undertakes environmental research and development and consultancy work for industry, community organizations, research institutions, and government agencies, among others in line with country's Vision 2030. Technologies and innovations developed by the division are transferred through business incubation, development of pilot plants, and provision of common manufacturing facilities.

#				
10	The Federation of Kenya Employers (FKE)	The national umbrella organization representing employers' interests and advocates an environment favorable to enterprise competiveness, sustainability and job creation.		There are over 2,500 members in the Federation, representing interests from such diverse sectors as: mining; timber harvesting and agriculture; manufacturing; banking and other financial services; retail, distribution and wholesale trade; civil engineering and building construction.
11	The Commission for University Education (CUE)	Established by an Act of Parliament (Universities Act Chapter 210B)	Policy Higher Education standards, curricular approval, issue charter	A corporate body charged to make better provisions for the advancement of quality university education in Kenya. CUE does not formulate policies but primarily deals with accreditation of programs such as those in climate change, pollution and occupational health. CUE uses experts to accredit programs i.e. peer review system. In case of environmental matters CUE uses experts from NEMA and other specialists at university level.
12	Chartered Public and private Universities	Moi, Egerton, Maseno, Nairobi, Technical, Strathmore, USIU, Kenyatta, Masinde Muliro, Masai Mara, Jaramogi OO,		Research, policy (through faculty participation in various committees and fora), consultancy, community service, training (graduate and undergraduate levels)

Appendix 8: Table 8: Universities and their training profile

Serial #	Institution	Courses offered / Services offered to training institutions
1.	University of Nairobi	1. Bachelor of Science in Environmental Geoscience 2. Master of Arts (Climatology) 3. Master of Arts (Environmental Planning and Management) 4. Master of Arts in Environmental Policy 5. Master of Science in Climate Change 6. Doctor of Philosophy in Agro-Ecosystems and Environment **7.** Doctor of Philosophy in Climate Change and Adaptation
2.	Maasai Mara University	1. MSc Environmental studies; 2. PhD Environmental Studies 3. BSc Environmental Studies
3.	University of Kabianga	1. BSc Environmental health (EVH); 2. BSc Environmental sciences (ES); 3. MSc Environmental Health; 4. MSc Environmental Sciences; 5. Diploma in Environmental health; 6. PhD at in EVH, ES
4.	Moi University	1. Bachelor of Science (Environmental Health) 2. Master of Science in Occupational Health & Safety 3. Doctor of Philosophy in Energy Studies
5.	Kenyatta University	1. Bachelor of Environmental Planning and Management 2. Bachelor of Environmental Science (Resource Conservation) 3. Master of Environmental Science (Resource Conservation) 4. Master of Environmental Planning and Management 5. Master of Environmental Studies (Agroforestry and Rural Development) **6.** Masters in Environmental Studies (Climate Change And Sustainability) 7. Bachelor of Environmental Health 8. Bachelor of Occupational Safety and Health 9. Master of Environmental Health 10. Master of Occupational Safety and Health

6.	Jomo Kenyatta University of Science and Technology	1. Bachelor of Science in Occupation health 2. Master of Science Degree in Occupational Safety in Health 3. Master of Science in Energy Technology
7.	Egerton University	1. Bachelor of Science in Natural Resources Management 2. Bachelor of Science in Environmental Science 3. Master of Science in Water Resources and Environmental Management 4. Doctor of Philosophy in Environmental and Occupational Health **5.** Doctor of Philosophy in Environmental Science
8.	Maseno University	1. Bachelor of Science in Climate Change & Development. 2. Bachelor of Science in Earth Science 3. Bachelor of Science in Environmental Science 4. Bachelor of Science in Geography & Natural Resource Management **5.** Doctor of Philosophy in Environmental Science
9.	Masinde Muliro University	1. Bachelor of Science in Bio-resources Management and Conservation 2. Bachelor of Science in Environmental Management and Conservation **3.** Master of Science in climate change, adaptation and Sustainable Development
10.	Mount Kenya University	MSc in conservation.

Appendix 9: Table 9: Key research institutions in Kenya and their research mandates

	INSTITUTION	RESEARCH FUNCTION	LINK WITH OTHER RESEARCH INSTITUTIONS
1.	UNIVERSITIES	Conduct research on all the three thematic areas of climate change, occupational health and air pollution Some research done by graduate students on self support basis; Some research done by faculty supported by local institutions such as NACOSTI, LVEMP etc; Most research supported by foreign agencies, are time bound, but again limited by spatial and temporal challenges;	Have collaborations and linkages a part of their policies Many are making inroads in this, and may help reduce the current research challenges
2.	NATIONAL ENVIRONMENT MANAGEMENT AUTHORITY (NEMA)	Coordinates all the environmental activities of the country Has developed an environment research policy and priority which is already guiding other research institutions on research focus;	Has a strong link with universities through the Regional Centre of Expertise (RCE) network in which they work together to find solutions to community environmental challenges.
3.	DIRECTORATE OF OCCUPATIONAL SAFETY AND HEALTH SERVICES (DOSHS)	The Occupational Health Division in DOSHS undertakes occupational health surveillance in workplaces. Monitors and supervises the activities of the designated health practitioners who carry out medical examination of workers. The Directorate's occupational hygiene and occupational health divisions are responsible for analytical and assessment work related to the determination of workers' exposure to various occupational hazards.	Partners closely in joint research with universities offering academic programs in occupational health and safety FKE also offers training on OSH, with trainers who are approved by the DOSHS. COTU trains its affiliated members, although the provision is based on the availability of funds.
4.	AMREF (AFRICAN MEDICAL RESEARCH FOUNDATION)	Carries out projects on various areas of health and environment, there is very limited use of research or recommendations from local universities. Provide university students with attachment opportunities to be able to develop field skills.	

5.	KENYA BUREAU OF STANDARDS (KEBS)	Has published a number of Kenya standards covering the environment sector. Is already partnering with the flower council of Kenya to evaluate the carbon footprint of their products, with a view to assessing their acceptability in global markets both in the short run and in the long run.	The regional Standards bureaus of which KEBS is part are already engaging on high profile research on carbon footprint of products (CFP), alongside universities (Maasai Mara and Egerton Universities) and the industry e.g. the flower council of Kenya. This is to enable the industry meet the long term trade standards for their products.
6.	NATIONAL COMMISSION FOR SCIENCE, TECHNOLOGY AND INNOVATION (NACOSTI)	The NACOSTI Act facilitates the promotion, coordination and regulation of the progress of Science, Technology and Innovation (ST&I) in the country. Research is therefore its core business. It offers financial and technical support to research partners such as universities. NACOSTI seeks to promote and mainstream R & D activities that are coordinated and all inclusive that aim to provide clean, secure and sustainable environment.	Research Approval & Surveillance Department (RASD) is the department under NACOSTI that licenses and monitors authorized research as per the Science Technology and Innovation Act of 2013.
7.	KIPPRA (KENYA INSTITTUTE FOR PUBLIC POLICY RESEARCH AND ANALYSIS)	Conducts objective research and analysis on public policy issues with the goal of providing advice to policy makers. Collects and analyses relevant data on public policy and disseminates its research findings to a wide range of stakeholders through workshops/conferences. Undertakes contracted public policy research and analysis for the government and clients from the private sector.	At the national level, KIPPRA collaborates with various government ministries, the Institute of Policy Analysis and Research, Teemed Institute of Agricultural Policy, Institute of Economic Affairs, Institute of Development Studies (University of Nairobi), and School of Economics at the University of Nairobi. Also collaborates with the local public and private institutions in areas of policy formulation, implementation and evaluation.
8.	KENYA MEDICAL RESEARCH INSTITUTE (KEMRI)	The Health Safety and Environmental Policy of KEMRI governs the use of hazardous materials in research and/or teaching.	
9.	KENYA INDUSTRIAL RESEARCH AND DEVELOPMENT INSTITUTE	The Environment Management Division is one of the research, technology and innovation (RTI) departments in KIRDI. The division undertakes environmental research and development and consultancy work for industry, community organizations, research institutions, and government agencies, among others in line with country's Vision 2030. Technologies and innovations developed by the division are transferred through business incubation,	Production of Biogas from the water hyacinth plant; Production of activated carbon from bagasse and water hyacinth plant; Utilization of activated carbon for waste water treatment; Development of gas stove utilizing agricultural wastes; Development of improved stoves for clean development mechanism (CDM) projects;

		development of pilot plants, and provision of common manufacturing facilities. The division undertakes research and development and technology transfer and training.	Evaluation and Development of Occupation Safety and Health systems in metal fabrication small and micro enterprises (SME's); Mapping out and calculating carbon footprints; The Institute has signed Memoranda of Understanding (MOU) with some universities to establish a framework of partnership between KIRDI and the other parties. Has established strong linkages and works in collaboration with several national, regional and international organizations and agencies.
10.	CUE (COMMISSION FOR UNIVERSITY EDUCATION)	The CUE Strategic Plan 2010-2015 has been developed to operationalize on going reforms in university education articulated by the Sessional Paper No. 1 of 2005 on Education, Training and Research and the Public Universities Inspection Board (PUIB) Report (2006). The reforms are anchored in the National Strategy for the Development of University Education (2008) and provide opportunities for the Commission to re-position and re-invent itself. The mandate of the Commission is to ensure the maintenance of standards, quality and relevance in all aspects of university education, training and research.	The Commission is a key partner of higher education institutions with the same goal namely; the creation of competitive world class universities that contribute to social and economic development through teaching research and innovation. CUE functions to advise and make recommendations to the Government on matters relating to university education and research requiring the consideration of the Government. To plan and provide for the financial needs of university education and research needs of universities; Coordination of government funded research projects by the universities
11.	THE KENYA METEOROLOGICAL SERVICE	Provision of meteorological and climatological services to all relevant sectors; Organization and administration of surface and upper air meteorological observations within its area of responsibility and the publication of climatological data; • Co-ordination of research in meteorology and climatology including co-operation with other authorities in all aspects of applied meteorological research, and the maintenance of the National Meteorological Library; Evolvement of suitable training programs in all fields of meteorology and other related scientific subjects which are relevant to the development of Kenya and other countries that participate in the Department's training activities.	Works closely with Universities with climatology departments and courses for practical exposure and research.

ISBN: 9966-7205-9-6

AIR POLLUTION, OCCUPATIONAL SAFETY, HEALTH, AND CLIMATE CHANGE:
FINDINGS, RESEARCH NEEDS, AND POLICY IMPLICATIONS
Establishing a GEO Health Hub Platform for Eastern Africa

www.ingramcontent.com/pod-product-compliance
Lightning Source LLC
Chambersburg PA
CBHW052016230326
41598CB00078B/3519